SpringerBriefs in Water Science and Technology

W0037571

More information about this series at http://www.springer.com/series/11214

Samiha A.H. Ouda · Abd El-Hafeez Zohry
Mostafa Morsy

Cropping Pattern Modification to Overcome Abiotic Stresses

Water, Salinity and Climate

 Springer

Samiha A.H. Ouda
Water Requirements and Field Irrigation
  Research Department
Soils, Water and Environment Research
  Institute, Agricultural Research Center
Giza
Egypt

Mostafa Morsy
Astronomy and Meteorology Department,
  Faculty of Science (Boys)
Al-Azhar University
Cairo
Egypt

Abd El-Hafeez Zohry
Crops Intensification Research Department
Field Crops Research Institute, Agricultural
  Research Center
Giza
Egypt

ISSN 2194-7244   ISSN 2194-7252 (electronic)
SpringerBriefs in Water Science and Technology
ISBN 978-3-319-69879-3   ISBN 978-3-319-69880-9   (eBook)
https://doi.org/10.1007/978-3-319-69880-9

Library of Congress Control Number: 2017955645

Printed on acid-free paper

This Springer imprint is published by Springer Nature
The registered company is Springer International Publishing AG
The registered company address is: Gewerbestrasse 11, 6330 Cham, Switzerland

Contents

Chapter 1
Introductory Overview of the Projected Distress

Samiha A.H. Ouda and Abd El-Hafeez Zohry

Abstract The urgency of feeding the growing population in Egypt, while combating soil pollution, salinization, and desertification has given plant and soil productivity research vital importance. Furthermore, high population increase will increase food production–consumption gaps, as well as food insecurity. Egypt suffers from several food gaps, namely wheat and maize. In addition, there is a gap in legumes, sugar, and oil crops. Because water resources in Egypt are becoming limited and scarce, Egypt will face a problem to allocate water to agriculture to maintain food security. Moreover, soil salinity is an enormous problem for agriculture under irrigation. In addition, the abiotic stress thatclimate change will cause, i.e., water and heat stress can disturb physical and chemical processes in crops. Moreover, water requirements will increase for most of the cultivated crops. Consequently, cropping pattern in Egypt will be highly affected by these anticipated stresses. Therefore, cropping pattern should be adjusted to combat these negative effects on cultivated area and food production in Egypt. This study is set to be implemented on the Nile Delta and Valley for governorates irrigated using surface irrigation from the Nile River, which is called old lands, as well as the areas on the edges of these governorates (new land and its soil is sandy). These areas are irrigated with irrigation systems, namely sprinkler or drip systems, depending on the cultivated crop. The recorded cropping pattern in Egypt in 2014/15 growing season was used as a base for comparison in this study. Furthermore, this study deals with how national cropping pattern can be modify to overcome abiotic stresses, such as food insecurity, water scarcity, induced salinity and climate change, to reduce their negative effects on the cultivated area and consequently food production. Thus, different cropping patterns were suggested and evaluated to achieve that.

Keywords Food gaps in Egypt · Water scarcity · Salinity stress · Climate change impacts

© The Author(s) 2018   1
S.A.H Ouda et al., *Cropping Pattern Modification to Overcome Abiotic Stresses*, SpringerBriefs in Water Science and Technology,
https://doi.org/10.1007/978-3-319-69880-9_1

Introduction

The beginning of twenty-first century is marked by global scarcity of water resources, environmental pollution, and increased salinization of soil and water. Increasing human population and reduction in land available for cultivation are two threats for agricultural sustainability (Shahbaz and Ashraf 2013). The increase in population has put pressure on land to increase productivity per unit area, unit time, and for unit resource used. Food shortages in many parts of the world, as well as the threat of insufficient supplies in the near future, continues to stimulate more intensive agricultural investigation in a search for more productive alternatives. Increasing food production requires putting more land under cultivation (Godfray et al. 2010). Furthermore, various environmental stresses, namely high winds, extreme temperatures, soil salinity, drought and flood have affected the production and cultivation of agricultural crops, among these, soil salinity is one of the most devastating environmental stresses, which causes major reductions in cultivated land area, crop productivity, and quality (Yamaguchi and Blumwald 2005). Thus, most of the extra food needs must come from higher production from land already being farmed. The major part of this increase is likely to come from increasing the number of crops produced per year on a given land, which offers potential not only to increase food production but also decrease land degradation (Godfray et al. 2010).

Cropping pattern is the yearly sequence and spatial arrangement of crops or of crops and fallow on a given area. It indicates the proportion of area under different crops at a point of time (Madari and Shekadar 2015). Cropping pattern should provide enough food for the family, fodder for cattle, and generate sufficient cash income for domestic and cultivation expenses. There are different definitions for cropping pattern existed in the literature. Cropping pattern is evolved based on climate, water availability, and soil for efficient use of available natural resources. The key atmospheric elements that impact crops are solar radiation, air temperature, humidity, and precipitation. The daily variability of these elements across the landscape can be described as weather. Weather extremes at critical periods of a crop's development can have dramatic influences on productivity of crops (Valipour 2017). The long-term average temperature and humidity, as well as the total solar radiation and precipitation over a crop's growing season can be described as the climate. It is the climate that, in the absence of any weather extremes, determines the realized yields for a given region (Hollinger and Angel 2014). Within cropping pattern, other crops systems can be implemented, such as inter-cropping and succession cropping. An intercropping system is two or more crops share the same piece of land for part, or for all, of their growing season (Eskandari et al. 2009). To ensure the optimum productivity in an intercropping system, the peak periods of growth of the two crops should not coincide, so that one quick maturing crop completes its life cycle before the main period of growth of the other crop starts (Parsons 2003). On the other hand, Gallaher (2009) defined succession cropping as two or more crops are grown in succession on the same land per year.

These forms are generally known as double cropping, triple cropping, and quadruple cropping and ratoon cropping. In Egypt, it is very common to implement double cropping and it is possibly to implement triple cropping. In fact, triple intercropping proved to improve soil fertility and increase farmer net income (Sheha et al. 2014; Zohry et al. 2017a, b).

The urgency of feeding the growing population in Egypt, while combating soil pollution, salinization, and desertification has given plant and soil productivity research vital importance. Furthermore, high population increase will increase food production–consumption gaps, as well as food insecurity. Egypt suffers from several food gaps, namely wheat and maize. In addition, there is a gap in legumes, sugar, and oil crops.

Water resources in Egypt are becoming limited and scarce. It was reported by the Ministry of Irrigation and Water Resources in Egypt in 2014 that Egypt will reach the threshold of absolute scarcity, namely 500 m^3/capita/year in 2025. Egypt received 55.5 BCM/year of Nile water. This limited share of the Nile water is not expected to increase in the future. Taking into account population growth and the expected negative effect of climate change, Egypt will face a problem to allocate water to agriculture to maintain food security.

Soil salinity is an enormous problem for agriculture under irrigation. In the hot and dry regions of the world, the soils are frequently saline with low agricultural potential. In these areas, most crops are grown under irrigation, and to exacerbate the problem, inadequate irrigation management leads to secondary salinization that affects 20% of irrigated land worldwide (Glick 2007). Irrigated agriculture is a major human activity, which often leads to secondary salinization of land and water resources in arid and semiarid conditions (Patel et al. 2011). Saline soils distribution is closely related to environmental factors such as climate, geology, geochemical, and hydrological conditions (Pimentel 2006). The cropping pattern in Egypt is somewhat adjusted to soil condition. In the northern part of the Nile Delta where soil salinity is somewhat high, crop rotation includes rice and cotton as the main summer crops and wheat and clover as the main winter crops. All of these crops have proved to be salt-tolerant or semi-tolerant (Ouda et al. 2016b).

Climate change poses unprecedented challenges to agriculture because of the sensitivity of agricultural productivity to changing climate (IPCC 2007). The abiotic stress that climate change will cause, i.e., water and heat stress can disturb physical and chemical processes in crops. Drought and temperature extremes can cause extensive economic loss to agriculture (Peng et al. 2004). Thus, agriculture will highly suffer from these expected effects of climate change. Moreover, under climate change effect in 2030, Ouda et al. (2016a) stated that water requirements will increase for most of the cultivated crops. Furthermore, intensified evaporation will increase the hazard of salt accumulation in the soil (Kazi et al. 2002). Climate change will affect crop production as changes in soil, air temperature, and rainfall affect the ability of crops to reach maturity and their potential harvest (Karmakar et al. 2016). Furthermore, the formation of soil affected by climate, which is one of the most important factors affecting with important implications for their development, use and management perspective with reference to soil structure, stability, top soil water holding capacity, nutrient availability, and erosion (Manchanda and Garg 2008). Further indirect effects correspond to changes in growth rates or water-use efficiencies, through sea-level rise, through climate-induced decrease or increase in vegetative cover or anthropogenic intervention (Rustad et al. 2000).

Thus, cropping pattern in Egypt will be highly affected by these anticipated stresses. Thus, cropping pattern should be adjusted to combat these negative effects on cultivated area and food production in Egypt. Munns (2002) indicated that farming systems can change to incorporate perennials in rotation with annual crops (phase farming), in mixed plantings (alley farming, intercropping), or in site-specific plantings (precision farming).

This study is set to be implemented on the Nile Delta and Valley for governorates irrigated using surface irrigation from the Nile River, which is called old lands. The areas on the edges of these governorates are also included in the analysis. These areas are called new lands and it is mainly sandy soil. These areas are irrigated with irrigation systems, namely sprinkler or drip systems, depending on the cultivated crop. The irrigated area with underground water was excluded from this analysis. Figure 1.1 showed a map of Egypt.

The recorded cropping pattern in Egypt in 2014/15 growing season is presented in Table 1.1 and it was used as a base for comparison in this study. The table included the most important crops from economic point of view and less economic crops were gathered in one category, namely other winter/summer crops. The table

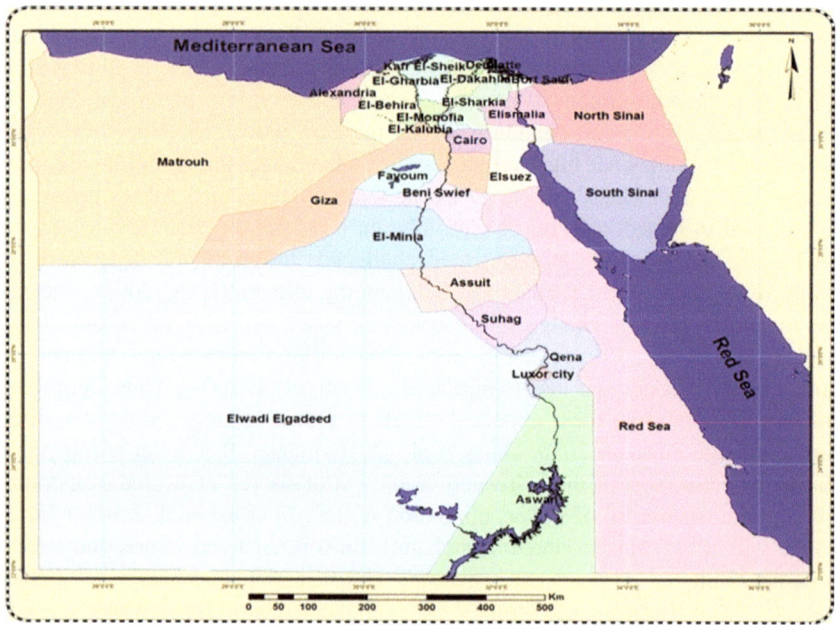

Fig. 1.1 A map for Egypt governorates

Table 1.1 National cropping pattern in Egypt in 2014/15 growing season and its water requirements

	Cultivated area (ha)	Water requirements (m³)
Winter crops		
Wheat	1,354,844	8,825,826,373
Faba bean	34,418	178,997,045
Clover	624,740	5,948,296,442
Sugar beet	231,193	1,955,202,088
Others	1,000,908	4,790,663,377
Total	3,246,104	21,698,985,325
Summer crops		
Cotton	100,349	1,264,924,487
Rice	506,249	5,896,134,782
Maize	938,329	9,259,942,383
Sugarcane	134,656	4,271,214,589
Fruit trees	510,776	9,789,068,856
Others	805,677	10,143,385,640
Total	2,996,036	40,624,670,737
Grand total	6,242,139	62,323,656,062

showed that the highest cultivated area in the winter was assigned to wheat, followed by clover. Similarly, the highest cultivated area in the summer was assigned to maize followed by rice. The perennial crops, namely sugarcane and fruit trees cultivated on 134.7 and 510.8 thousand hectares, respectively. The total cultivated area was 6.2 million hectares in old and new lands on national level.

Water requirements for the studied cropping pattern was calculated for the same year and presented in Table 1.1.

Thus, this study deals with how national cropping pattern can be modify to overcome abiotic stresses, such as food insecurity, water scarcity, induced salinity, and climate change, to reduce their negative effects on the cultivated area and consequently food production. Thus, different cropping pattern were suggested and evaluated to achieve that.

Conclusion

This chapter provided an introductory overview on the projected distresses that will threaten food availability in Egypt. Food availability to the growing population, water scarcity and salinization of the agricultural lands and negative effects of climate change are the most stressful environmental factors that negatively affect the cultivated cropping pattern and food production. The current cropping pattern in Egypt does not supply the needed foods for the growing population. Taking into consideration the previously mentioned stressful factors, modification in the cropping pattern can be done to minimize these distresses. Lastly, this study can also be a framework for other developing countries to increase their food security.

References

Eskandari H, Ghanbari A, Javanmard A (2009) Intercropping of cereals and legumes for forage production. Notulae Scientia Biologicae 1:07–13

Gallaher RN (2009) Management of agricultural forestry and fisheries enterprises. Vol. I: Multiple Cropping Systems Encyclopedia of Life Support Systems (EOLSS)

Godfray HCJ, Beddington JR, Crute IR, Haddad L, Lawrence D, Muir JF, Pretty J, Robinson S, Thomas SM, Toulmin C (2010) Food security: the challenge of feeding 9 billion people. Science 327:812–818

Glick BR (2007) Promotion of plant growth by bacterial ACC deaminase. Crit Rev Plant Sci 26 (2007):227–242

Hollinger SE, Angel JR (2014) Weather and crops. In: Illinois agronomy handbook. University of Illinois. USA. pp 1–12

IPCC Intergovernmental Panel on Climate Change (2007) Intergovernmental panel on climate change fourth assessment report: climate change 2007. Synthesis Report. World Meteorological Organization, Geneva, Switzerland

Karmakar R, Das I, Dutta D, Rakshit A (2016) Potential effects of climate change on soil properties: a review. Sci Int 4:51–73

Kazi BR, Oad FC, Jamro GH, Jamali LA, Oad NL (2002) Effect of water stress on the growth, yield, and oil content of sunflower. Pak J Appl Sci 2(5):550–552

Madari DM, Shekadar SI (2015) Impact of irrigation on cropping pattern and production with special reference to Vijaour district. Golden Res Thoughts 4(8):320–325

Manchanda G, Garg N (2008) Salinity and its effects on the functional biology of legumes. Acta Physiol Plant 30:595–618

Munns R (2002) Comparative physiology of salt and water stress. Plant Cell Environ 25:239–250

Ouda S, Abd El-Latif K, Khalil F (2016a) Water requirements for major crops. In: Major crops and water scarcity in Egypt. Springer Publishing House, Berlin, pp 25–31

Ouda S, Zohry AA, Khalifa H (2016b) Combating deterioration in salt-affected soil in egypt by crop rotations. In: Management of climate induced drought and water scarcity in Egypt: unconventional solutions. Springer Publishing House, Berlin. ISBN: 978-3-319-33659-6

Parsons MJ (2003) Successful intercropping with sugarcane. Proc S Afr Sug Technol Ass 77: 77–98

Patel BB, Patel BhB, Dave RS (2011) Studies on infiltration of saline–alkali soils of several parts of Mehsana and Patan districts of North Gujarat. J Appl Technol Environ Sanitation 1(1): 87–92

Peng S, Huang J, Sheehy JE, Laza RC, Visperas RM, Zhong X, Centeno GS, Khush GS, Cassman KG (2004) Rice yields decline with higher night temperature from global warming. Natl Acad Sci 101:9971–9975

Pimentel D (2006) Soil erosion: a food and environmental threat. Environ Dev Sustain 8:119–137

Rustad LE, Huntington TG, Boone RD (2000) Controls on soil respiration: implications for climate change. Biogeochemistry 48:1–6

Shahbaz M, Ashraf M (2013) Improving salinity tolerance in cereals. Crit Rev Plant Sci 32: 237–249

Sheha AM, Ahmed NR, Abou-Elela AM (2014) Effect of crop sequence and nitrogen levels on rice productivity. Ann Agri Sci 52(4): 451–460

Valipour M (2017) Analysis of potential evapotranspiration using limited weather data. Appl Water Sci 7:187–197. doi:10.1007/s13201-014-0234-2

Yamaguchi T, Blumwald E (2005) Developing salt-tolerant crop plants: challenges and opportunities. Trends Plant Sci 10(12):615–620

Zohry AA, Abbady K, El-Mazz AE, Ahmed H (2017a) Maximizing land productivity by diversified cropping systems with different nitrogen types. Acta Agriculturae Slovenica (under review)

Zohry AA, Ouda S, Hamd-Alla W, Shalaby E (2017b) Evaluation of different crop sequences for wheat and maize in sandy soil. Acta agriculturae Slovenica 109(2):383

Chapter 2
Water Requirements for Prevailing Cropping Pattern

Samiha A.H. Ouda and Abd El-Hafeez Zohry

Abstract The objective of this chapter was to calculate water requirements for the prevailing cropping pattern in the five agroclimatic zones of Egypt. Weather data were collected for 2014/15 growing seasons to calculate water requirements for the studied cropping pattern for the five agroclimatic zones. BISm model was used to calculate ETo. The planting and harvest dates for 19 important crops that existed in the cropping pattern was determined. The date of each growth stage and the values of crop coefficients for each of the studied crops, as well as its water consumptive use were then calculated by the model. These calculations will help in the determination of water requirements for each of the studied crops in the level of each agroclimatic zone and on the national level.

Keywords Agroclimatic zones of Egypt · ETo · Crops coefficients · Crops water requirements

Introduction

The actual water resources currently available for use in Egypt are 55.5 BCM/year from the Nile River, and 1.3 BCM/year effective rainfalls on the northern strip of the Nile Delta, 1.3 BCM/year of nonrenewable ground water in the western desert and Sinai, whereas water requirements for different sectors are in the order of 76.4 BCM/year. The gap between the needs and availability of water is about 20 BCM/year, which is overcome by recycling agricultural drainage water. Of this amount agriculture consumes 85% of the total water resources, namely 62.3 BCM/year (Ministry of Irrigation and Water Resources 2014).

Furthermore, on-farm irrigation systems are low efficient coupled with poor irrigation management and surface irrigation is the major system applied to 83% of the old cultivated lands in the Nile Delta and Valley (Abou Zeid 2002). Application efficiency of surface irrigation in Egypt on farm level is 60%, which endure large losses in the applied irrigation water to drainage canals. Thus, it is highly imperative to use it most judiciously to ensure sustainable agriculture development and productivity. This, in turn requires knowledge of crop water requirements in various agroclimatic zones of Egypt. A superior water management program seeks to provide an optimal balance of water and air in the soil, which allows full expression of genetic potential in plants. The differences among poor, average, and record crop yields generally can be attributed to the amount and timing of the soil's water supplies (Cooke 2012).

Thus, the objective of this chapter was to calculate water requirements for the prevailing cropping pattern in the five agroclimatic zones of Egypt.

Agroclimatic Zones of Egypt

The Basic Irrigation Scheduling (BIS) application was used to calculate water requirements for the cultivated crops. The model is written using MS Excel to help people plan irrigation management of crops. The BIS application calculates evapotranspiration (ETo) using the Penman–Monteith equation (Monteith 1965) as presented in the United Nations FAO Irrigation and Drainage Paper (FAO 56) by Allen et al. (1998) and using the Hargreaves–Samani equation (1985). If only temperature data are input, Hargreaves–Samani equation is used. For ETo calculations, the station latitude and elevation must also be input. After calculating daily means per month, a cubic spline curve fitting subroutine is used to estimate daily ETo rates for the entire year. The model requires sowing and harvest dates as input and irrigation frequency to calculate crop kc. The model also account for water depletion from root zone. Therefore, it requires to input total water holding capacity and available water. The model then determines the time when irrigation needs to be applied and the required amounts (Snyder et al. 2004).

ETo values for 30 years were calculated by BISm model (Snyder et al. 2004) in 17 governorates in the Nile Delta and Valley. ETo values was averaged over 10, 20, and 30 years from 1985 to 2014 and graphed together in Fig. 2.1 for comparison purpose. The figure showed that lowest value for ETo were found using 30-year average, followed by 20-year average then 10-year average in all studied sites, except Alexandria governorate. These results implied that weather elements are in continuous rising over the years.

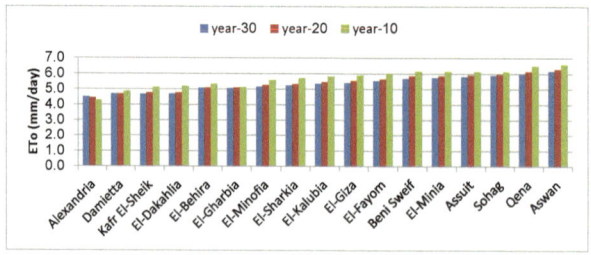

Fig. 2.1 Comparison between values of ETo averaged over 10, 20, and 30 years

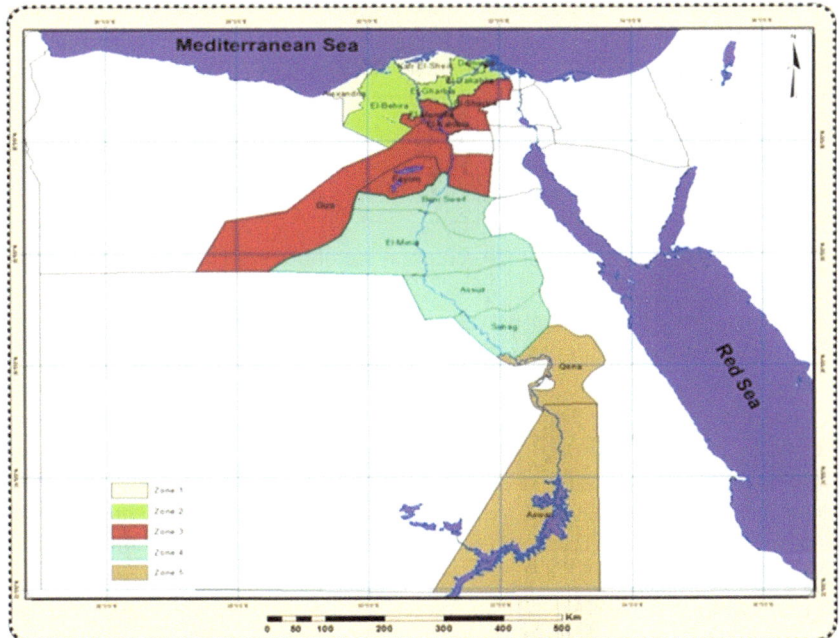

Fig. 2.2 Map of agroclimatic zones of Egypt using 10-year of ETo values

Noreldin et al. (2016), Ouda and Noreldin (2017) used ETo values for these three time periods, namely 30, 20, and 10 year from 1985 to 2014 to develop agroclimatic zones for Egypt. In that methodology, monthly means of weather data for 10-year were calculated for each governorate. Analysis of variance was used and the means was separated and ranked using least significant difference test ($LSD_{0.05}$). Zoning using 10-year values of ETo resulted in five agroclimatic zones only and higher values of ETo in each zone, compared to 20-year and 30-year ETo values. Figure 2.2 showed the five agroclimatic zones developed by Ouda and Noreldin (2017).

Water Requirements for the Prevailing Cropping Pattern

Hess (2005) stated that the ICID-CIID (2000) defined crop water requirements as the total water needed for evapotranspiration, from planting to harvest for a given crop in a specific climate regime, when adequate soil water is maintained by rainfall and/or irrigation so that it does not limit plant growth and crop yield. Whereas, USDA, Soil Conservation Service (1993) defined it as the amount of water required to compensate the evapotranspiration loss from a cropped field.

Calculation of water requirements depends on ETo. ETo is the total amount lost from the field by both soil evaporation and plant transpiration (Gardner et al. 1985). Accurate estimation of ETo is important factor to attain prop water management. Earlier studies compared different ETo equations for their accuracy and revealed that Penman–Monteith equation is the most accurate because of its detailed theoretical base and its accommodation of small time periods (Valipour 2017). It was found that air temperature and solar radiation contributed most to the temporal variation of ETo in the upper reaches, as well as solar radiation and wind speed were the determining factors for the temporal variation of ETo in the middle-lower reaches (Zhao et al. 2015).

Water consumptive use can be calculated by multiplying ETo with crop coefficient (kc). Water consumptive use accounts for variations in weather and offers a measure of the "evaporative demand" of the atmosphere. Whereas, crop coefficient (kc) account for the difference between ETc and ET (Snyder et al. 2004). The kc takes into account the relationship between atmosphere, crop physiology, and agricultural practices (Lascano 2000). Water consumptive use is affected by planting date, which reflects the weather of a certain site. Planting date also could affect growth pattern of a crop and consequently affects the period of growth stages, the value of kc and growth period (Gardner et al. 1985). As a crop canopy develops, the ratio of transpiration to ETo increases until most of the ETo comes from transpiration and evapotranspiration is a minor component. This occurs because light interception by the foliage increases until most light is intercepted before it reaches the soil (Snyder et al. 2004). Therefore, crop coefficients for field crops generally increase until the canopy ground cover reaches about 75% and the light interception is near 80% (Snyder et al. 2004). The accurate calculation of crop kc for each growth stage is an important component for accurate calculation of water requirements (Shideed et al. 1995).

To calculate water requirements for the studied cropping pattern, weather data were collected for 2014/15 growing seasons for the five agroclimatic zones. BISm model was used to calculate ETo. The kc for each of the studied crops and its water consumptive use were then calculated. Table 2.1 showed planting and harvest dates for the selected crops. It worth noting that there is a range of time a crop can be cultivated in. However, for calculation purpose, a certain date was assigned for planting.

In the first agroclimatic zone, the dates of each growth stage are presented in Table 2.2. All the strategic crops are cultivated in the first agroclimatic zone, except sugarcane, where it is only cultivated in south Egypt.

Table 2.3 presented the value of each crop coefficients and water consumptive use for the studied crops in the first agroclimatic zone.

The most important cultivated crops in the second agroclimatic zone are presented in Table 2.4, as well as the date of each growth stage.

Values of crop coefficient for the studied crops and its water requirements in the second agro climatic zone are presented in Table 2.5. Several studies were done to calculate water consumptive use for crops grown in this zone. The measured values of water consumptive use for wheat in 2009/10 and 2010/11 were 355–384 mm

Table 2.1 Planting and harvest dates of selected crops in the prevailing cropping pattern

Crop	Planting date	Harvest date
Wheat	15 Nov	18 Apr
Faba bean	25 Oct	25 Mar
Clover	15 Oct	1 Apr
Onion	15 Nov	15 May
Tomato	1 Oct	1 Mar
Potato	1 Nov	1 Feb
Sugar beet	15 Oct	12 Apr
Cotton	15 Mar	1 Sep
Rice	15 May	16 Sep
Maize	15 May	1 Sep
Soybean	15 May	25 Aug
Sunflower	15 May	15 Aug
Tomato	1 May	1 Sep
Citrus*	15 Feb	14 Feb
Olive*	15 Feb	14 Feb
Grape*	15 Feb	14 Feb

*End of the agricultural year

Table 2.2 The dates of each growth stage for the studied crops in the first agroclimatic zone in 2015

Crop	A–B	C	D	E
Wheat	15 Nov–16 Dec	23 Jan	11 Mar	18 Apr
Faba bean	25 Oct–30 Nov	24 Dec	12 Mar	25 Mar
Clover	15 Oct–26 Oct	4 Dec	15 Mar	1 Apr
Onion	15 Nov–3 Dec	1 Jan	31 Mar	15 May
Tomato	1 Oct–8 Nov	16 Dec	30 Jan	1 Mar
Potato	1 Nov–19 Nov	12 Dec	12 Jan	1 Feb
Sugar beet	15 Oct–11 Nov	4 Jan	8 Mar	12 Apr
Cotton	15 Mar–9 Apr	26 Apr	7 Aug	1 Sep
Rice	15 May–14 Jun	30 Jun	30 Aug	16 Sep
Maize	15 May–6 Jun	30 Jun	5 Aug	1 Sep
Soybean	15 May–4 Jun	30 Jun	5 Aug	25 Aug
Sunflower	15 May–2 Jun	25 Jun	28 Jul	15 Aug
Tomato	1 May–1 Jun	2 Jul	8 Aug	1 Sep
Citrus	15 Feb	15 Jun	17 Oct	14 Feb
Olive	15 Feb	15 Jun	17 Oct	14 Feb
Grape	15 Feb	28 Apr	20 Sep	14 Feb

Table 2.3 Values of crop coefficients for the cultivated crops in the first agroclimatic zone and its water consumptive use

Crop	Crop coefficient (kc)				Water consumptive use (mm)
	A–B	C	D	E	
Wheat	0.34	1.09	1.09	0.21	363
Faba bean	0.31	0.96	0.92	0.22	338
Clover	0.29	1.15	1.15	1.15	526
Onion	0.33	1.20	1.20	0.54	615
Tomato	0.28	1.10	1.10	0.64	313
Potato	0.31	1.09	1.09	0.68	199
Sugar beet	0.29	1.15	1.15	0.95	508
Cotton	0.31	0.95	0.95	0.49	725
Rice	0.40	1.02	1.02	0.78	667
Maize	0.26	1.04	1.04	0.58	535
Soybean	0.27	1.10	1.10	0.40	530
Sunflower	0.27	1.09	1.09	0.38	474
Potato	0.25	1.10	1.10	0.69	422
Tomato	0.27	1.10	1.10	0.65	611
Citrus	0.30	1.0	1.0	1.0	1412
Olive	0.30	0.8	0.8	0.8	1155
Grape	0.37	0.8	0.8	0.35	874

Table 2.4 The dates of each growth stage for the studied crops in the second agroclimatic zone

Crop	A–B	C	D	E
Wheat	15 Nov–16 Dec	23 Jan	11 Mar	18 Apr
Faba bean	25 Oct–30 Nov	24 Dec	12 Mar	25 Mar
Clover	15 Oct–26 Oct	4 Dec	15 Mar	1 Apr
Onion	15 Nov–3 Dec	1 Jan	31 Mar	15 May
Tomato	1 Oct–8 Nov	16 Dec	30 Jan	1 Mar
Potato	1 Nov–19 Nov	12 Dec	12 Jan	1 Feb
Sugar beet	15 Oct–11 Nov	4 Jan	8 Mar	12 Apr
Cotton	15 Mar–9 Apr	26 Apr	7 Aug	1 Sep
Rice	15 May–14 Jun	30 Jun	30 Aug	16 Sep
Maize	15 May–6 Jun	30 Jun	5 Aug	1 Sep
Soybean	15 May–4 Jun	30 Jun	5 Aug	25 Aug
Sunflower	15 May–2 Jun	25 Jun	28 Jul	15 Aug
Potato	1 Aug–25 Aug	24 Sep	2 Nov	28 Nov
Tomato	1 May–1 Jun	2 Jul	8 Aug	1 Sep
Citrus	15 Feb	15 Jun	17 Oct	14 Feb
Olive	15 Feb	15 Jun	17 Oct	14 Feb
Grape	15 Feb	28 Apr	20 Sep	14 Feb

Table 2.5 Values of crop coefficients for the cultivated crops in the second agroclimatic zone and water consumptive use

Crop	Crop coefficients (kc)				Water consumptive use (mm)
	A–B	C	D	E	
Wheat	0.33	1.09	1.09	0.20	385
Faba bean	0.30	0.96	0.96	0.21	355
Clover	0.28	1.15	1.15	1.15	558
Onion	0.32	1.20	1.20	0.54	663
Tomato	0.27	1.10	1.10	0.64	343
Potato	0.30	1.09	1.09	0.68	206
Sugar beet	0.28	1.15	1.15	0.95	541
Cotton	0.30	0.95	0.95	0.49	792
Rice	0.38	1.02	1.02	0.78	722
Maize	0.25	1.04	1.04	0.58	579
Soybean	0.25	1.10	1.10	0.40	572
Sunflower	0.25	1.09	1.09	0.37	509
Potato	0.23	1.10	1.10	0.69	451
Tomato	0.26	1.10	1.10	0.65	661
Citrus	0.30	1.00	1.00	1.00	1532
Olive	0.30	0.80	0.80	0.80	1253
Grape	0.37	0.80	0.80	0.35	955

(Taha 2012). In another experiment in the same zone, the measured values of water consumptive use for wheat grown in five growing seasons from 2001/02 season until 2005/06 season was around 334–349 mm (Ouda et al. 2008). Regarding maize, its water consumptive use was 478–507 mm in 2009 and 2010 growing seasons (Taha 2012). Table 2.5 showed that water consumptive use in was 385 mm for wheat and 579 mm for maize, which imply the increasing trend in weather data through the past decades.

Table 2.6 showed the date of growth stages in the most important cultivated crops in the third agroclimatic zone in Egypt.

Table 2.7 indicated that the value of kc for the selected crops was slightly different in the third agroclimatic zone, compared with the first agroclimatic zone. The table also showed that water consumptive use values in 2014/15 growing season were 409, 577, 597, and 530 mm, for wheat, sugar beet, maize, and sunflower, respectively. Khalil et al. (2007) stated that water consumptive use for wheat was 381–401 mm in 2005/06 and 2006/07 growing seasons. Measured values of water consumptive use for sugar beet was between 534 and 565 mm in 2010/11 growing season (personal communication). With respect to maize, Khalil and Mohamed (2006) indicated that maize water consumptive use was 541–546 mm, whereas sunflower water consumptive use was 531–536 mm (Khalil 2007). The differences in water consumptive value can be attributed to the effect of climate change on increasing these values in the third agroclimatic zone.

Table 2.6 The dates of each growth stage for the studied crops in the third agroclimatic zone

Crop	A–B	C	D	E
Wheat	15 Nov–16 Dec	23 Jan	11 Mar	18 Apr
Faba bean	25 Oct–30 Nov	24 Dec	12 Mar	25 Mar
Clover	15 Oct–26 Oct	4 Dec	15 Mar	1 Apr
Onion	15 Nov–3 Dec	1 Jan	31 Mar	15 May
Tomato	1 Oct–8 Nov	16 Dec	30 Jan	1 Mar
Potato	1 Nov–19 Nov	12 Dec	12 Jan	1 Feb
Sugar beet	15 Oct–11 Nov	4 Jan	8 Mar	12 Apr
Cotton	15 Mar–9 Apr	26 Apr	7 Aug	1 Sep
Rice	15 May–14 Jun	30 Jun	30 Aug	16 Sep
Maize	15 May–6 Jun	30 Jun	5 Aug	1 Sep
Soybean	15 May–4 Jun	30 Jun	5 Aug	25 Aug
Sunflower	15 May–2 Jun	25 Jun	28 Jul	15 Aug
Potato	1 Aug–25 Aug	24 Sep	2 Nov	28 Nov
Tomato	1 May–1 Jun	2 Jul	8 Aug	1 Sep
Citrus	15 Feb	15 Jun	17 Oct	14 Feb
Olive	15 Feb	15 Jun	17 Oct	14 Feb
Grape	15 Feb	17 May	15 Nov	14 Feb

Table 2.7 Value of crop coefficients for the cultivated crops in the third agroclimatic zone and water consumptive use

Crop	Crop coefficient (kc)				Water consumptive use (mm)
	A–B	C	D	E	
Wheat	0.31	1.08	1.08	0.19	409
Faba bean	0.28	0.96	0.96	0.20	375
Clover	0.27	1.15	1.15	1.15	598
Onion	0.30	1.20	1.20	0.54	707
Tomato	0.26	1.10	1.10	0.64	364
Potato	0.28	1.09	1.09	0.68	216
Sugar beet	0.27	1.15	1.15	0.95	577
Cotton	0.28	0.95	0.95	0.49	830
Rice	0.36	1.02	1.02	0.78	740
Maize	0.24	1.04	1.04	0.58	597
Soybean	0.24	1.10	1.10	0.40	592
Sunflower	0.24	1.09	1.09	0.37	530
Potato	0.23	1.10	1.10	0.69	473
Tomato	0.24	1.10	1.10	0.65	679
Citrus	0.29	0.8	0.8	0.8	1607
Olive	0.35	0.8	0.8	0.35	1314
Grape	0.29	0.95	0.95	0.95	996

Table 2.8 The dates of each growth stage for the studied crops in the fourth agroclimatic zone

Crop	A–B	C	D	E
Wheat	15 Nov–16 Dec	23 Jan	11 Mar	18 Apr
Faba bean	25 Oct–30 Nov	24 Dec	12 Mar	25 Mar
Clover	15 Oct–26 Oct	4 Dec	15 Mar	1 Apr
Onion	15 Nov–3 Dec	1 Jan	31 Mar	15 May
Tomato	1 Oct–8 Nov	16 Dec	30 Jan	1 Mar
Potato	1 Nov–19 Nov	12 Dec	12 Jan	1 Feb
Sugar beet	15 Oct–11 Nov	4 Jan	8 Mar	12 Apr
Cotton	15 Mar–9 Apr	26 Apr	7 Aug	1 Sep
Maize	15 May–6 Jun	30 Jun	5 Aug	1 Sep
Soybean	15 May–4 Jun	30 Jun	5 Aug	25 Aug
Sunflower	15 May–2 Jun	25 Jun	28 Jul	15 Aug
Potato	1 Aug–25 Aug	24 Sep	2 Nov	28 Nov
Tomato	1 May–1 Jun	2 Jul	8 Aug	1 Sep
Sugarcane	15 Mar	22 Jun	3 Sep	1 Mar
Citrus	15 Feb	15 Jun	17 Oct	14 Feb
Olive	15 Feb	15 Jun	17 Oct	14 Feb
Grape	15 Feb	17 May	15 Nov	14 Feb

For the fourth agroclimatic zone, the date of crops growth stages for the studied crops is presented in Table 2.8. In this zone, rice is not planted, as it is prohibited by law. Furthermore, sugarcane is cultivated in this zone.

The effect of weather of the agroclimatic zone is shown in this zone resulted in differences in the values of kc for the studied crops, compared to the other mentioned zones (Table 2.9). Furthermore, Ouda et al. (2010a) stated that the value of faba bean water consumptive use was 355–366 mm in 2000/01 growing season (Ouda et al. 2010b), whereas it was 392 mm in 2014/15 season. The value of cotton water consumptive use was 876–899 mm in 2000 and 2001 growing seasons. El-Sayed (2016) indicated that water consumptive use for cotton in the fourth agroclimatic zone was 1005–1056 mm in 2012 and 2013 respectively, whereas it was 905 mm in 2014/15 season (Table 2.9).

The date of growth stages in the fifth agroclimatic zones is presented in Table 2.10, where rice is prohibited for cultivation there. Furthermore, the weather of this zone is not suitable for cotton and sugar beet.

Table 2.11 showed the values of crop coefficients and water consumptive use of the studied crops in the fifth agroclimatic zone.

Water requirements for selected crops at each zone can be calculated under surface irrigation in the old lands of the Nile Delta and Valley using 60% application. In the new lands, water requirements for selected crops at each zone can be calculated under irrigation systems, namely sprinkler with 75% application efficiency and drip systems with 85% application efficiency.

Table 2.9 Values of crop coefficients for the cultivated crops in the fourth agroclimatic zone and water consumptive use

Crop	Crop coefficient (kc)				Water consumptive use (mm)
	A–B	C	D	E	
Wheat	0.31	1.08	1.08	0.18	431
Faba bean	0.28	0.96	0.96	0.20	392
Clover	0.25	1.15	1.15	1.15	623
Onion	0.30	1.20	1.20	0.54	750
Tomato	0.24	1.10	1.10	0.64	378
Potato	0.28	1.09	1.09	0.68	222
Sugar beet	0.25	1.15	1.15	0.95	604
Cotton	0.28	0.95	0.95	0.49	905
Maize	0.23	1.04	1.04	0.58	643
Soybean	0.23	1.10	1.10	0.40	638
Sunflower	0.23	1.09	1.09	0.37	574
Potato	0.22	1.10	1.10	0.69	524
Tomato	0.23	1.10	1.10	0.65	735
Sugarcane	0.21	1.2	1.2	0.62	1971
Citrus	0.27	0.8	0.8	0.8	1735
Olive	0.34	0.8	0.8	0.35	1416
Grape	0.27	0.95	0.95	0.95	1082

Table 2.10 The dates of each growth stage for the studied crops in the fifth agroclimatic zone

Crop	A–B	C	D	E
Wheat	15 Nov–16 Dec	23 Jan	11 Mar	18 Apr
Faba bean	25 Oct–30 Nov	24 Dec	12 Mar	25 Mar
Clover	15 Oct–26 Oct	4 Dec	15 Mar	1 Apr
Onion	15 Nov–3 Dec	1 Jan	31 Mar	15 May
Tomato	1 Oct–8 Nov	16 Dec	30 Jan	1 Mar
Potato	1 Nov–19 Nov	12 Dec	12 Jan	1 Feb
Maize	15 May–6 Jun	30 Jun	5 Aug	1 Sep
Soybean	15 May–4 Jun	30 Jun	5 Aug	25 Aug
Sunflower	15 May–2 Jun	25 Jun	28 Jul	15 Aug
Potato	1 Aug–25 Aug	24 Sep	2 Nov	28 Nov
Tomato	1 May–1 Jun	2 Jul	8 Aug	1 Sep
Sugarcane	15 Feb	18 Apr	21 Oct	14 Feb
Citrus	15 Feb	15 Jun	17 Oct	14 Feb
Olive	15 Feb	15 Jun	17 Oct	14 Feb
Grape	15 Feb	17 May	15 Nov	14 Feb

Table 2.11 Values of crop coefficients for the cultivated crops in the fifth agroclimatic zone and water consumptive use

Crop	Crop coefficient (kc)				Water consumptive use (mm)
	A–B	C	D	E	
Wheat	0.30	1.08	1.08	0.18	451
Faba bean	0.27	0.96	0.96	0.19	413
Clover	0.25	1.15	1.15	1.15	659
Onion	0.29	1.20	1.20	0.54	787
Tomato	0.24	1.10	1.10	0.64	400
Potato	0.27	1.09	1.09	0.68	239
Maize	0.21	1.03	1.03	0.58	645
Soybean	0.22	1.10	1.10	0.40	643
Sunflower	0.22	1.09	1.09	0.37	577
Potato	0.22	1.10	1.10	0.69	538
Tomato	0.23	1.10	1.10	0.65	743
Sugarcane	0.21	1.20	1.20	0.62	2028
Citrus	0.27	0.80	0.80	0.80	1792
Olive	0.33	0.80	0.80	0.35	1463
Grape	0.27	0.95	0.95	0.95	1097

Conclusion

Climate variability in the past 30 years in Egypt increased ETo values, which affected water requirements for the growing crops in the prevailing cropping pattern. To improve irrigation water management in Egypt, agroclimatic zones was developed using the latest 10-year values of ETo and five agroclimatic zones. BISm model was used to calculate crops kc values, the date of each stage for the selected crops and water consumptive use in each of the five agroclimatic zones. These calculations will help in the determination of water requirements for the cultivated crops on the level of each agroclimatic zone and on the national level.

References

Abou Zeid K (2002) Egypt and the world water goals, Egypt statement in the world summit for sustainable development and beyond, Johannesburg, South Africa

Allen RG, Pereira LS, Raes D, Smith M (1998) Crop evapotranspiration: guideline for computing crop water requirements. FAO No. 56

Cooke R (2012) Water Management. In: Illinois agronomy handbook. University of Illinois, USA, pp 143–152

El-Sayed AEM (2016) Response of Egyptian cotton to alternative systems of irrigation and different rates of splitting of NPK under two planting dates. Ph.D. thesis. Assuit University

Gardner FP, Pearce RB, Mitchell RL (1985) Physiology of crop plants. Iowa State University Press, Ames, USA

Hargreaves GH, Samani ZA (1985) Reference crop evapotranspiration from temperature. Transaction of ASAE 1(2):96–99

Hess T (2005) Crop water requirements, water and agriculture, water for agriculture. WCA infoNET, USA

ICID-CIID (2000) Multilingual technical dictionary on irrigation and drainage.—CD Version September 2000. International Commission on Irrigation and Drainage, New Dehli

Khalil FAF (2007) Effect of some agricultural practices on productivity and water use efficiency for sunflower. Minufiya J Agric Res 32(1):283–296

Khalil FAF, Mohamed SGA (2006) Studies on the inter-relationship among irrigation and maize varieties on yield and water relations using some statistical procedures. Anal Agric Sci Moshtohor 44(1):393–406

Khalil FA, El-Shaarawy GA, Hassan YM (2007) Irrigation scheduling for some wheat cultivars through pan evaporation norms and its effect on growth, yield and water use efficiency. Fayoum J Agric Res Dev 21(1):222–233

Lascano RJ (2000) A general system to measure and calculate daily crop water use. Agron J 92:821–832

Ministry of Irrigation and Water Resources (2014) Water scarcity in Egypt: the urgent need for regional cooperation among the Nile Basin countries. Technical report

Monteith JL (1965) Evaporation and environment. In: Fogg GE (ed) Symposium of the society for experimental biology: the state and movement of water in living organisms, vol 19. Academic Press, Inc., NY, pp 205–234

Noreldin T, Ouda S, Amer A (2016) Agro-climatic zoning in the Nile Delta and Valley to improve water management. J Water Land Dev Water Land Dev 31(X–XII):113–117

Ouda S, Noreldin T (2017) Evapotranspiration data to determine agro-climatic zones in Egypt. J Water Land Dev 32(I–III):79–86

Ouda SA, AbouElenin R, Shreif MA (2010a) Increasing water productivity of faba bean grown under deficit irrigation at middle Egypt. In: Proceedings of the 14th international conference on water technology, Egypt

Ouda SA, Khalil FA, AbouElenin R, Shreif MA, Benli B, Qadir M (2008) Using yield-stress model in irrigation management for wheat grown in Egypt. J Appl Biol Sci 2(1):57–65

Ouda SA, Abd El-Baky H, AbouElenin RM, Shreif A (2010b) Simulation of the effect of irrigation water management on cotton yield at two locations in Egypt. In: Proceedings of the 14th international conference on water technology, Egypt

Shideed K, Oweis T, Gabr M, Osman M (1995) Assessing on-farm water use efficiency: a new approach. Ed. ICARDA/ESCWA, Aleppo, Syria, 86 pp

Snyder RL, Orang M, Bali K, Eching S (2004) Basic Irrigation Scheduling (BIS). http://www.waterplan.water.ca.gov/landwateruse/wateruse/Ag/CUP/Californi/Climate_Data_010804.xls

Taha A (2012) Effect of climate change on maize and wheat grown under fertigation treatments in newly reclaimed soil. Ph.D. Thesis, Tanta University, Egypt

USDA, Soil Conservation Service (1993) Irrigation water requirements. National Engineering Handbook NEH, Part 623, Chapter 2, National technical information service

Valipour M (2017) Analysis of potential evapotranspiration using limited weather data. Appl Water Sci 7:187–197. doi:10.1007/s13201-014-0234-2

Zhao J, Xu Z, Zuo D, Wang X (2015) Temporal variations of reference evapotranspiration and its sensitivity to meteorological factors in Heihe River Basin, China. Water Sci Eng 8(1):1–8

Chapter 3
Prevailing Cropping Pattern

Abd El-Hafeez Zohry and Samiha A.H. Ouda

Abstract The objective of this chapter was to assess the cultivated area in the cropping pattern recorded in 2014/15 growing seasons in the five agroclimatic zones in Egypt, as well as its water requirements. The distribution of the five main crops (wheat, maize, clover, cotton, and sugarcane) in the five agroclimatic zones was assessed, with respect to its cultivated area, as well as its applied irrigation amounts in 2014/15 cropping pattern. In addition to these five main crops, other crops are also cultivated and its water requirements were calculated. The total cultivated area was found to be around 6.3 million hectares, required around 62.3 billion cubic meters of irrigation water. The cultivated area of wheat represented that highest area in the winter season, followed by clover and sugar beet. Whereas, in the summer season, maize had the highest cultivate area followed by fruit trees and rice. The prevailing cropping pattern resulted in food gaps in wheat, maize, faba bean, oil crops, sugar crops, and summer forage crops. Thus, it needs to be modified to reduce these food gaps and increase food security.

Keywords Main crops in Egypt · Other cultivated crops in Egypt · Agroclimatic zones of Egypt · Food gaps in Egypt

Introduction

Cropping pattern is the agricultural land devoted to different crops in a region or state or country at a particular point of time. The cropping pattern of a region is an outcome of a long-term agricultural practices, social customs and traditions, physical conditions, and historical factors (http://nagahistory.wordpress.com/2014/03/15/land-use-and-agriculture/). Moreover, cropping pattern obtained in any particular agricultural area is generally the outcome of trials and adjustment, in respect of farm enterprise and practices (Madari and Shekadar 2015). Since the beginning of the past century, cropping pattern in Egypt has been composed of six major crops, namely cotton, maize, wheat, rice, clover, and sugarcane.

© The Author(s) 2018
S.A.H Ouda et al., *Cropping Pattern Modification to Overcome Abiotic Stresses*, SpringerBriefs in Water Science and Technology,
https://doi.org/10.1007/978-3-319-69880-9_3

Egyptian agriculture possesses certain features that make it unique among other agricultural systems all over the world. Such uniqueness is the outcome of the combined effects of these features. The old lands in the Nile Delta and Valley are the main growing areas in Egypt that are characterized by complex-year long cropping pattern. The Northern Delta region is characterized by high salinity, especially near the Mediterranean coast and lakes. The richest crop production area is the Mid-Delta region due to the high quality of soil. Upper Egypt is characterized by arid weather; thus, certain types of crops are being grown there. Reclaimed agricultural lands in the desert are characterized by having advanced technology; yet their constraints arise from the low fertility.

Thus, the objective of this chapter was to assess the cultivated area in the cropping pattern recorded in 2014/15 growing seasons in the five agroclimatic zones in Egypt, as well as its water requirements.

The Prevailing Cropping Pattern

The cropping pattern in Egypt depends on five main crops, namely wheat, maize, clover, cotton, and sugarcane. Wheat and maize for flour production, clover for feed; cotton for fiber and edible oil and sugarcane for sugar production. Other crops exist in the Egyptian cropping pattern, such as cereals, legumes, fibers, forages, vegetables, and fruit crops. Nowadays, Egypt has production–consumption gap in wheat and maize. There are production–consumption gaps in feed production, especially in summer. The cultivated areas of cotton were reduced, as a result of high production cost. Furthermore, there is a production–consumption gap in sugar, partially compensated by the expansion in sugar beet crop. Other food gaps exist in Egypt, namely faba bean production–consumption gap, edible oil production–consumption gap.

For more detailed assessment for the current cropping pattern, it was distributed on the five agroclimatic zones developed by Ouda and Noreldin (2017) (Table 3.1). In this zoning, Egypt was divided on the basis of the value of evapotranspiration of each governorate.

First Agroclimatic Zone

This zone is composed of two governorates, namely Alexandria and Kafr El-Sheik (Table 3.1). Its area is equal to 5767 km^2, both located on the Mediterranean Sea. All the major crops are cultivated in this zone in both old and new lands, except sugarcane. Furthermore, soybean was not cultivated in the new lands of this zone. The total cropped area in old and new land in the winter and summer growing

Table 3.1 The agroclimatic zones of Egypt

Zone number	Governorate	ETo (mm/day)
First agroclimatic zone	Alexandria	4.279
	Kafr El-Sheik	4.852
Second agroclimatic zone	Demiatte	5.123
	El-Dakahlia	5.344
	El-Behira	5.192
	El-Gharbia	5.125
Third agroclimatic zone	El-Monofia	5.800
	El-Sharkia	5.869
	El-Kalubia	5.964
	Giza	5.701
	Fayom	5.587
Fourth agroclimatic zone	Beni Sweif	6.139
	El-Minia	6.140
	Assuit	6.122
	Sohag	6.127
Fifth agroclimatic zone	Qena	6.480
	Aswan	6.600

season of 2014/15 was 569,413 and 142,952 ha, respectively. In the winter season and in the old lands, wheat ranked first with respect to its cultivated area, followed by sugar beet and clover. In the new lands and in the same growing season, still wheat ranked first with respect to highest cultivated area, whereas clover had a larger area than sugar beet. In the summer season, rice had the highest cultivated area, followed by cotton and maize in the old lands. In the new lands, tomato ranked first with respect to highest cultivated area, followed by maize (Table 3.2). The term "others" in Table 3.2 refers to the cultivated area of a large number of legumes, fibers, forages, vegetables, and fruit crops gathered together in this category.

The evapotranspiration value (ETo) for this zone is between 4.279 and 4.852 mm/day (Table 3.1), which is the lowest values in Egypt. The total water requirements to irrigate this area in both winter and summer seasons were 4,374,270,550, and 481,661,343 m^3 for the old and new cultivated lands, respectively (Table 3.2).

Table 3.3 indicated that the total cultivated area was 712,005 ha of both old and new lands required the application of 4,855,931,893 m^3 of irrigation water. Wheat and rice had the highest cultivated area in the winter and summer growing seasons, respectively. Furthermore, both crops consumed the highest value of irrigation water.

Table 3.2 Cultivated area and its water requirements in old and new cultivated area in the first agroclimatic zone

| | Old cultivated lands | | New cultivated lands | |
	Area (ha)	Water requirements (m³)	Area (ha)	Water requirements (m³)
Winter crops				
Wheat	108,146	654,282,292	22,882	110,747,267
Faba bean	5696	32,086,976	1900	7,556,951
Clover	51,878	454,813,695	6627	46,477,944
Onion	18,153	139,474,910	72	518,529
Tomato	2140	11,162,206	312	1,149,201
· Potato	3776	12,525,821	1256	2,941,103
Sugar beet	65,115	551,325,177	2980	17,807,392
Others	6487	46,704,000	80,857	91,812,000
Total	261,390	1,902,375,078	116,886	279,010,388
Summer crops				
Cotton	31,009	374,694,097	372	3,173,652
Rice	102,139	1,135,442,438	1714	13,447,897
Maize	30,658	273,370,139	5700	35,873,848
Soybean	430	3,794,653	0	0
Sunflower	32	250,167	86	480,971
Potato	217	1,523,889	3213	15,953,255
Tomato	9520	96,941,090	10,795	77,599,995
Sugarcane	0	0	0	0
Fruit trees	5860	112,805,000	1770	24,056,838
Others	128,159	473,074,000	2055	32,064,500
Total	308,023	2,471,895,472	25,706	202,650,956
Grand total	569,413	4,374,270,550	142,592	481,661,343

Second Agroclimatic Zone

This zone is composed of four governorates, namely Demiatte, El-Dakahlia, El-Behira, and El-Gharbia (Table 3.1). All these governorates located on the Mediterranean Sea, except El-Gharbia, where it is located in the middle of the Nile Delta. The total area of this zone is equal to 17,561 km². All the major crops are cultivated in this zone, except sugarcane in both old and new lands. The total cropped area in the winter and summer growing season of 2014/15 in the old and new land was 1,432,198 and 727,466 ha, respectively. Similar to the first agro-climatic zone, wheat ranked first with respect to highest cultivated area, however clover cultivated area was higher than sugar beet cultivated in the winter season and in the old lands. In the new lands and in the same growing season, still wheat ranked first with respect to highest cultivated area, whereas sugar beet had a larger

Table 3.3 Total cultivated area and total water requirements used in the first agroclimatic zone

	Area (ha)	Water requirements (m^3)
Winter crops		
Wheat	131,028	765,029,558
Faba bean	7597	39,643,927
Clover	58,505	501,291,640
Onion	18,225	139,993,439
Tomato	2452	12,311,407
Potato	5033	15,466,924
Sugar beet	68,094	569,132,569
Others	87,344	138,516,000
Total	378,276	2,181,385,465
Summer crops		
Cotton	31,381	377,867,749
Rice	103,853	1,148,890,335
Maize	36,358	309,243,987
Soybean	430	3,794,653
Sunflower	118	731,137
Potato	3430	17,477,144
Tomato	20,315	174,541,085
Sugarcane	0	0
Fruit trees	7630	136,861,838
Others	130,215	505,138,500
Total	333,729	2,674,546,428
Grand total	712,005	4,855,931,893

area, compared to clover. Furthermore, in the summer season, rice had the highest cultivated area, followed by maize and cotton in the old lands. In the new lands, fruit trees ranked first with respect to highest cultivated area, followed by maize (Table 3.4).

The value of ETo for this zone is between 5.123 and 5.344 mm/day (Table 3.1). The total water requirements to irrigate this area in both winter and summer seasons were 14,484,332,350 and 6,684,874,323 m^3 for the old and new cultivated lands, respectively (Table 3.4).

The total cultivated area in the second agroclimatic zone was 2,159,665 ha of both old and new lands. This large area required the application of 21,169,206,674 m^3 of irrigation water. Similar to the first agroclimatic zone, wheat and rice had the highest cultivated area in the winter and summer growing seasons, respectively. Furthermore, wheat consumed the highest value of irrigation water in the winter season, whereas fruit trees consumed the highest value of irrigation water in the summer season (Table 3.5).

Table 3.4 Cultivated area and its water requirements in old and new cultivated area in the second agroclimatic zone

	Old cultivated lands		New cultivated lands	
	Area (ha)	Water requirements (m³)	Area (ha)	Water requirements (m³)
Winter crops				
Wheat	323,818	2,077,942,245	78,373	402,312,167
Faba bean	8000	47,336,000	10,672	44,571,642
Clover	216,421	2,012,713,750	24,380	181,384,100
Onion	48,878	399,173,056	8137	63,469,250
Tomato	3278	18,739,850	14,694	59,293,603
Potato	40,967	140,638,567	16,238	39,353,069
Sugar beet	55,534	500,751,581	33,456	212,939,191
Others	36,849	274,154,700	162,033	843,865,694
Total	733,745	5,471,449,748	347,982	1,847,188,716
Summer crops				
Cotton	37,316	492,574,500	6513	60,685,059
Rice	280,003	3,369,373,444	24,766	210,367,441
Maize	170,062	1,641,099,104	43,672	297,481,118
Soybean	451	4,301,917	396	2,663,725
Sunflower	1303	11,049,542	2606	15,606,838
Potato	23,606	177,440,313	5760	30,564,093
Tomato	10,867	119,719,035	28,368	220,605,510
Sugarcane	0	0	0	0
Fruit trees	56,715	1,184,389,549	214,730	3,165,372,824
Others	118,130	2,012,935,200	52,673	834,339,000
Total	698,453	9,012,882,603	379,485	4,837,685,608
Grand total	1,432,198	14,484,332,350	727,466	6,684,874,323

Third Agroclimatic Zone

Five governorates are located in the third agroclimatic zone, namely El-Monofia, El-Sharkia, El-Kalubia, Giza, and Fayom (Table 3.1). The first three governorates are located in Nile Delta and the fourth and the fifth governorates are located in Middle Egypt. The total area of this zone is equal to 41,855 km². All the major crops are cultivated in this zone, except sugarcane in both old and new lands. The total cropped area in old and new land in the winter and summer growing season of 2014/15 was 1,513,430 and 370,238 ha, respectively. In the winter season and in the old lands, wheat ranked first with respect to highest cultivated area, followed by clover. In the new lands and in the same growing season, still wheat ranked first with respect to highest cultivated area, followed by sugar beet. In the summer

Table 3.5 Total cultivated area and total water requirements used in the second agroclimatic zone

	Area (ha)	Water requirements (m^3)
Winter crops		
Wheat	402,191	2,480,254,412
Faba bean	18,672	91,907,642
Clover	240,800	2,194,097,850
Onion	57,015	462,642,306
Tomato	17,972	78,033,453
Potato	57,205	179,991,635
Sugar beet	88,990	713,690,772
Others	198,881	1,118,020,394
Total	1,081,727	7,318,638,463
Summer crops		
Cotton	43,829	553,259,559
Rice	304,770	3,579,740,886
Maize	213,734	1,938,580,222
Soybean	847	6,965,642
Sunflower	3909	26,656,380
Potato	29,367	208,004,406
Tomato	39,235	340,324,545
Sugarcane	0	0
Fruit trees	271,445	4,349,762,372
Others	170,803	2,847,274,200
Total	1,077,938	13,850,568,211
Grand total	2,159,665	21,169,206,674

season, maize had the highest cultivated area, followed by fruit trees in the old lands. In the new lands, fruit trees ranked first with respect to highest cultivated area, followed by tomato (Table 3.6).

The value of ETo for this zone is between 5.587 and 5.964 mm/day (Table 3.1). The total water requirements to irrigate this area in both winter and summer seasons were 15,100,538,091 and 2,487,717,316 m^3 for the old and new cultivated lands, respectively (Table 3.6).

The total cultivated area in the third agroclimatic zone was 1,783,669 ha of both old and new lands. It required the application of 17,588,255,407 m^3 of irrigation water. Wheat and maize had the highest cultivated area in the winter and summer growing seasons, respectively. Furthermore, wheat consumed the highest value of irrigation water in the winter season, whereas fruit trees consumed the highest value of irrigation water in the summer season (Table 3.7).

Table 3.6 Cultivated area and its water requirements in old and new cultivated area in the third agroclimatic zone

	Old cultivated lands		New cultivated lands	
	Area (ha)	Water requirements (m³)	Area (ha)	Water requirements (m³)
Winter crops				
Wheat	357,174	2,434,853,454	31,966	174,322,617
Faba bean	3869	24,182,292	1235	5,450,368
Clover	192,927	1,922,904,240	8664	69,082,289
Onion	31,325	272,001,799	7750	64,461,765
Tomato	17,566	106,571,911	11,959	51,211,588
Potato	29,621	106,635,000	523	1,329,882
Sugar beet	34,219	329,085,726	13,372	90,772,848
Others	55,986	429,971,200	157,371	172,515,840
Total	722,686	5,626,205,620	232,841	629,147,197
Summer crops				
Cotton	17,745	245,472,500	4360	42,578,186
Rice	87,548	1,079,762,778	10,078	87,740,784
Maize	300,686	2,991,824,042	12,831	90,117,735
Soybean	125	1,237,444	0	0
Sunflower	858	7,578,264	625	3,897,059
Potato	19,093	150,512,542	417	2,318,627
Tomato	11,440	129,457,951	8276	66,109,304
Sugarcane	0	0	0	0
Fruit trees	106,566	2,333,791,750	88,890	1,374,134,824
Others	246,684	2,534,695,200	11,920	191,673,600
Total	790,744	9,474,332,471	137,397	1,858,570,120
Grand total	1,513,430	15,100,538,091	370,238	2,487,717,316

Fourth Agroclimatic Zone

Four governorates existed in this zone, namely Beni Sweif, El-Minia, El-Behira, Assuit, and Sohag (Table 3.1). An area of 80,181 km² is the total area of this zone. The first two governorates are located in the Middle Egypt and the second two governorates are located in the Upper Egypt. All the major crops are cultivated in this zone, except rice in the old lands and cotton and potato in the new lands. The total cropped area in the winter and summer growing season of 2014/15 in old and new lands was 1,001,775 and 130,812 ha, respectively. Wheat ranked first with respect to the highest cultivated area in the old and new lands. Furthermore, in the

Table 3.7 Total cultivated area and total water requirements used in the third agroclimatic zone

	Area (ha)	Water requirements (m^3)
Winter crops		
Wheat	389,140	2,609,176,070
Faba bean	5105	29,632,659
Clover	201,591	1,991,986,528
Onion	39,075	336,463,563
Tomato	29,525	157,783,499
Potato	30,144	107,964,882
Sugar beet	47,591	419,858,574
Others	213,357	602,487,040
Total	955,527	6,255,352,817
Summer crops		
Cotton	22,105	288,050,686
Rice	97,627	1,167,503,562
Maize	313,517	3,081,941,777
Soybean	125	1,237,444
Sunflower	1483	11,475,323
Potato	19,509	152,831,169
Tomato	19,715	195,567,255
Sugarcane	0	0
Fruit trees	195,456	3,707,926,574
Others	158,604	2,726,368,800
Total	828,141	11,332,902,590
Grand total	1,783,669	17,588,255,407

summer season, maize had the highest cultivated area in the old lands. In the new lands, fruit trees ranked first with respect to highest cultivated area, followed by maize (Table 3.8).

The value of ETo for this zone is between 6.122 and 6.140 mm/day (Table 3.1). The total water requirements to irrigate this area in both winter and summer seasons were 11,043,306,554 and 1,187,142,483 m^3 for the old and new cultivated lands, respectively (Table 3.8).

The total cultivated area in the fourth agroclimatic zone was 1,132,586 ha of both old and new lands. This area required the application of 12,230,449,036 m^3 of irrigation water. Wheat and maize had the highest cultivated area in the winter and summer growing seasons, respectively. Furthermore, wheat consumed the highest amount of irrigation water in the winter season, whereas maize consumed the highest amount of irrigation water in the summer season (Table 3.9).

Table 3.8 Cultivated area and its water requirements in old and new cultivated area in the fourth agroclimatic zone

| | Old cultivated lands | | New cultivated lands | |
	Area (ha)	Water requirements (m^3)	Area (ha)	Water requirements (m^3)
Winter crops				
Wheat	308,964	2,219,286,616	34,078	195,836,822
Faba bean	1,835	11,990,777	247	1,137,569
Clover	104,719	1,087,299,108	6169	51,241,750
Onion	23,467	214,723,813	8244	72,742,647
Tomato	4683	29,499,750	8806	39,161,912
Potato	9600	35,520,000	74	193,706
Sugar beet	21,646	217,908,604	4871	34,611,569
Others	17,529	138,831,000	17,755	98,436,030
Total	492,443	3,955,059,668	80,245	493,362,004
Summer crops				
Cotton	3033	45,746,493	0	0
Rice	0	0	0	0
Maize	327,472	3,509,404,694	13,165	99,589,353
Soybean	12,704	135,087,639	24	178,265
Sunflower	981	9,387,292	94	633,088
Potato	742	6,477,222	0	0
Tomato	5935	72,708,854	3732	32,267,941
Sugarcane	20,756	681,829,125	743	20,906,679
Fruit trees	35,353	834,338,667	13,987	233,008,353
Others	90,555	1,793,266,900	18,823	307,196,800
Total	509,332	7,088,246,886	50,567	693,780,479
Grand total	1,001,775	11,043,306,554	130,812	1,187,142,483

Fifth Agroclimatic Zone

Two governorates existed in this zone, namely Qena and Aswan (Table 3.1). An area of 75,934 km^2 is the total area of this zone. Both governorates are located in Upper Egypt. Several crops are not cultivated in this zone, namely sugar beet, cotton, rice, soybean, sunflower in the old lands and potato in the new lands. The total cropped area in the winter and summer growing season of 2014/15 in old and new lands was 221,494 and 302,164 ha, respectively. Wheat ranked first with respect to highest cultivated area in the old and new lands. Furthermore, in the

Table 3.9 Total cultivated area and total water requirements used in the fourth agroclimatic zone

	Area (ha)	Water requirements (m³)
Winter crops		
Wheat	343,042	2,415,123,438
Faba bean	2082	13,128,346
Clover	110,888	1,138,540,858
Onion	31,711	287,466,460
Tomato	13,489	68,661,662
Potato	9674	35,713,706
Sugar beet	26,517	252,520,173
Others	35,285	237,267,030
Total	572,688	4,448,421,672
Summer crops		
Cotton	3033	45,746,493
Rice	0	0
Maize	340,637	3,608,994,047
Soybean	12,728	135,265,904
Sunflower	1075	10,020,380
Potato	742	6,477,222
Tomato	9667	104,976,795
Sugarcane	21,498	702,735,804
Fruit trees	49,340	1,067,347,020
Others	109,379	2,100,463,700
Total	548,099	7,782,027,365
Grand total	1,120,786	12,230,449,036

summer season, sugarcane had the highest cultivated area in the old lands and new lands (Table 3.10).

The value of ETo for this zone is between 6.480 and 6.600 mm/day (Table 3.1), which is the highest value in Egypt. The total water requirements to irrigate this area in both winter and summer seasons were 4,858,572,257 and 2,023,515,380 m³ for the old and new cultivated lands, respectively (Table 3.10).

The total cultivated area in the fifth agroclimatic zone was 520,302 ha of both old and new lands. It required the application of 6,479,813,052 m³ of irrigation water. Wheat and sugarcane had the highest cultivated area in the winter and summer growing seasons, respectively. Furthermore, wheat consumed the highest amount of irrigation water in the winter season, whereas sugarcane consumed the highest amount of irrigation water in the summer season (Table 3.11).

Table 3.10 Cultivated area and its water requirements in old and new cultivated area in the fifth agroclimatic zone

	Old cultivated lands		New cultivated lands	
	Area (ha)	Water requirements (m^3)	Area (ha)	Water requirements (m^3)
Winter crops				
Wheat	47,667	358,313,465	41,777	251,217,022
Faba bean	225	1,545,807	738	3,587,431
Clover	9227	101,336,480	3729	32,766,944
Onion	672	6,474,403	1475	13,656,765
Tomato	1086	7,239,251	5650	26,590,196
Potato	3	11,617	227	637,333
Sugar beet	0	0	0	0
Others	2022	16,496,800	142,804	818,551,020
Total	60,901	491,417,823	196,400	1,147,006,712
Summer crops				
Cotton	0	0	0	0
Rice	0	0	0	0
Maize	25,061	269,408,438	9023	68,464,853
Soybean	0	0	0	0
Sunflower	0	0	0	0
Potato	63	560,417	0	0
Tomato	614	7,600,271	278	2,429,319
Sugarcane	98,492	3,329,032,417	14,666	424,890,143
Fruit trees	17,323	422,380,292	7680	132,186,353
Others	19,041	338,172,600	74,117	248,538,000
Total	160,593	4,367,154,433	105,763	876,508,667
Grand total	221,494	4,858,572,257	302,164	2,023,515,380

Adding the cultivated area of the five agroclimatic zones, as well as its water requirements resulted in total cultivated area 6,169,389 ha, requires 62,323,656,063 m^3 of irrigation water. Wheat represented that highest cultivated area in the winter season, followed by clover and sugar beet. Whereas, in the summer season, maize had the highest cultivate area followed by fruit trees and rice (Table 3.12).

Food gaps problems are occupying great importance in the light of the steady increase in population, so as the demand for major food commodities, and lack of resources (Hafez et al. 2011). Currently, there is a gap between production and consumption of cereal crops, namely wheat and maize were the gap is 49 and 47%,

Table 3.11 Total cultivated area and total water requirements used in the fifth agroclimatic zone

	Area (ha)	Water requirements (m³)
Winter crops		
Wheat	89,444	556,242,894
Faba bean	963	4,684,470
Clover	12,956	122,379,567
Onion	2147	18,371,220
Tomato	6736	30,871,941
Potato	230	592,217
Sugar beet	0	0
Others	145,410	762,044,599
Total	257,886	1,495,186,909
Summer crops		
Cotton	0	0
Rice	0	0
Maize	34,084	321,182,350
Soybean	0	0
Sunflower	0	0
Potato	63	532,732
Tomato	892	9,534,128
Sugarcane	113,158	3,568,478,785
Fruit trees	892	527,171,052
Others	113,845	557,727,096
Total	262,932	4,984,626,144
Grand total	520,818	6,479,813,052

respectively (Ouda and Zohry 2017; Zohry and Ouda 2017a). There is a gap in the production of oil crops around 93%, where the cultivated area of sunflower and soybean are very low (Field Crops Research Institute, Egypt). Furthermore, there is a gap in sugar crops production, around 30%, where the cultivated area of sugarcane and sugar beet is not enough to attain self-sufficiency of sugar (Field Crops Research Institute, Egypt). There is a gap in legume crops, especially faba been, where expansion in the sugar beet cultivated area was done using the cultivated area of faba bean resulted in 73% production–consumption gap (Zohry and Ouda 2017b). Lastly, there is a gap in forage crops production, especially in the summer around 90% (Field Crops Research Institute, Egypt).

Table 3.12 National cropping pattern in Egypt in 2014/15 growing season and its water requirements

	Cultivated area (ha)	Water requirements (m^3)
Winter crops		
Wheat	1,354,844	8,825,826,373
Faba bean	34,418	178,997,045
Clover	624,741	5,948,296,442
Onion	148,173	1,244,936,988
Tomato	70,173	347,661,962
Potato	102,285	339,729,364
Sugar beet	231,193	1,955,202,088
Others	680,278	2,858,335,063
Total	3,246,104	21,698,985,326
Summer crops		
Cotton	100,349	1,264,924,487
Rice	506,249	5,896,134,782
Maize	938,329	9,259,942,383
Soybean	14,130	147,263,643
Sunflower	6585	48,883,220
Potato	53,110	385,322,673
Tomato	89,825	824,943,808
Cowpea	516	1,032,000
Sugarcane	134,656	4,271,214,589
Fruit trees	524,763	9,789,068,856
Others	682,329	8,735,940,296
Total	3,050,840	40,624,670,737
Grand total	6,296,944	62,323,656,063

Conclusion

The cropping pattern in Egypt depends on five main crops, namely wheat, maize, clover, cotton, and sugarcane. The distribution of these crops in the five agroclimatic zones was assessed, with respect to its cultivated area, as well as its applied irrigation amounts in 2014/15 cropping pattern. In addition to these five main crops, other crops are also cultivated and its water requirements were calculated. The total cultivated area was found to be 6,296,944 ha, required 62,323,656,063 m^3 of irrigation water. Wheat represented that highest cultivated area in the winter season, followed by clover and sugar beet. Whereas, in the summer season, maize had the highest cultivate area followed by fruit trees and rice. The assessed cropping pattern resulted in food gaps in wheat, maize, faba bean, oil crops, sugar crops, and summer forage crops. Thus, it needs to be modified to reduce these food gaps and increase food security.

References

Hafez W, Abd El-Monsif H, Manaa M (2011) Food security in Egypt in 2030: Future scenarios Available at: http://www.idsc.gov.eg/IDSC/publication/View.aspx?ID=352

Madari DM, Shekadar SI (2015) Impact of irrigation on cropping pattern and productivity with special reference to Vijapur district. Golden Research Thoughts. 4(8):1–9

Ouda S, Noreldin T (2017) Evapotranspiration data to determine agro-climatic zones in Egypt. J Water Land Dev 32(I–III):79–86

Ouda S, Zohry AA (2017) Crops intensification to reduce wheat gap in Egypt. In: Future of food gaps in Egypt: obstacles and opportunities. Springer Publishing House, Berlin

Zohry AA, Ouda S (2017a) Increasing land and water productivities to reduce maize food gap. In: Future of food gaps in Egypt: obstacles and opportunities. Springer Publishing House, Berlin

Zohry AA, Ouda S (2017b) Solution for Faba bean production-consumption Gap. In: Future of food gaps in Egypt: obstacles and opportunities. Springer Publishing House, Berlin

Chapter 4
Cropping Pattern to Increase Food Security

Abd El-Hafeez Zohry and Samiha A.H. Ouda

Abstract The objective of this chapter is to quantify the effect of using poly-cropping on increasing food security in the five agro-climatic zones of Egypt. Our results indicated that food gaps in Egypt can be decreased by increasing poly-cropping, where intercropping techniques and cultivation of three crops per year were implemented. Using intercropping systems for wheat, faba bean, maize, sunflower, and cowpea can increase its cultivated area to high percentage, 20 and 22%, respectively. Whereas, faba bean, sunflower, and cowpea cultivated area can be increased to very high percentage, reaching 611, 5500 and 128,186%, respectively. These high percentages of increase were a result of implementing inter-cropping and cultivation of three crops per year. Our results also showed that there was no need to apply extra irrigation water to cultivate the middle crop between winter and summer, namely short season clover or sunflower because it can be obtained from cultivation of all the suggested cropping systems on raised beds. Thus, the proposed cropping pattern can increase the national cultivated area by 35%, compared to current cultivated area in 2014/15. Moreover, the increase in the cultivated area will not consume extra irrigation water.

Keywords Polycropping · Intercropping techniques · Cultivation of three crops per year · Food gaps in Egypt · Agro-climatic zones of Egypt

Introduction

The concept of food security is based on three main pillars, food availability, food accessibility, and food stability. Agriculture is the producer of food using the available natural resources, namely soil, water and weather resources. Increasing the efficiency of using these resources can increase food production and availability,

© The Author(s) 2018
S.A.H Ouda et al., *Cropping Pattern Modification to Overcome
Abiotic Stresses*, SpringerBriefs in Water Science and Technology,
https://doi.org/10.1007/978-3-319-69880-9_4

as well as reduce food insecurity. To help in meeting world's demand for food, multiple cropping must be implemented, namely intercropping techniques and successive cropping. Intercropping techniques can attain all the above advantages. Furthermore, successive cropping can have many benefits, such as improve and sustain soil fertility, as well as increase farmers' income (Sheha et al. 2014). Both are solution when land is limited, thus intensive cropping can fully utilize available water and labor when sufficient (Gallaher 2009).

Intercropping is the growing of two or more crops simultaneously on the same field (Gomez and Gomez 1983), where one crop shares its life cycle or part of it with another crop (Eskandari et al. 2009). Not only it can integrate the advantages of well-developed eco-agricultural techniques, but also can partly overcome their limitations (Wu and Wu 2014). Intercropping provides year-round ground cover, or at least for a longer period than monocultures, in order to protect the soil from desiccation and erosion (Gebru 2015). By growing more than one crop at a time in the same field, farmers maximize water use efficiency and maintain soil fertility (Tolera 2003), where greater root concentrations of the soil profile occur. Resource partitioning in plants may change when they are grown in intercropping system, where greater percentages of total dry matter and nutrients are allocated to harvestable portions of crops (Geno and Geno 2001). Intercropping of plants with different rooting patterns permits greater exploitation of a larger volume of soil and improves access to relatively immobile nutrients (Gebru 2015). As a result, intercropped plants tend to absorb more nutrients than those in monocultures. The increased crop production often observed in intercrops, compared to sole crops has been attributed to enhanced resource use (Szumigalski and Van-Acker 2008). Furthermore, the cumulative nutrient use efficiency of an intercropping system was in most cases higher than either of the sole crops (Chowdhury and Rosario 1994). Physiological complementarity can occur in intercropping composed of species that use C4 and C3 photosynthetic pathways (Geno and Geno 2001). Intercropping practice could modify the microclimate by reducing light intensity, air temperature, desiccating wind, and other climatic components (Gebru 2015). Thus, intercropping can be used to produce more crops in the cropping pattern without any using extra irrigation water, allowing reduction of food insecurity in Egypt. Because our water resources are limited, it confines our ability to cultivate new areas. Thus, polycropping (using multiple crops in the same space) can attain sustainable use of natural resources. It causes ecological balance, more utilization of resources, increases the quantity and quality, and reduces yield damage by pests, diseases, and weeds (Mazaheri et al. 2006).

Thus, the objective of this chapter is to quantify the effect of using polycropping on increasing food security in the five agro-climatic zones of Egypt. Figure 4.1 showed the five agro-climatic zones developed by Ouda and Noreldin (2017).

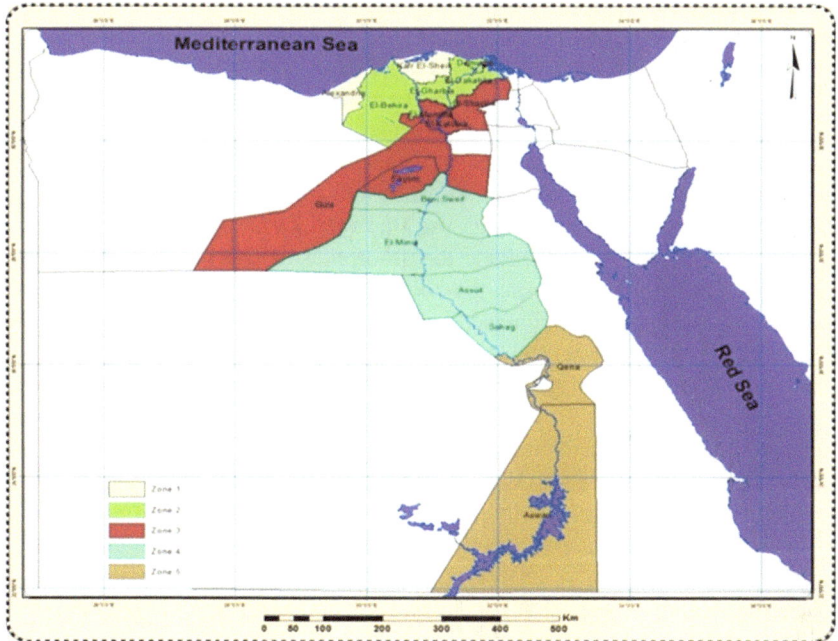

Fig. 4.1 Map of agro-climatic zones of Egypt using 10-year of ETo values

Increasing the Production of Major Crops Through Polycropping

Effect of Intercropping Techniques

Intercropping technique can increase the cultivated area of crops in Egypt, such as wheat, maize, sunflower, and summer forage crops. These crops have production–consumption gap. Intercropping should be implemented on raised beds, which save 20% of the applied irrigation water, compared to cultivation in basins or on narrow furrows (Abouelenein et al. 2010). Several investigations have been done in Egypt to highlight the roll of intercropping techniques in increasing production of these crops. In this context, sugar beet is a good candidate for intercropping, where two intercropping systems were successfully implemented in Egypt: wheat intercropping with sugar beet (Abou-Elela 2012) and faba bean intercropped with sugar beet (Abd El-Zaher and Gendy 2014). In both systems, sugar beet is the main crops, which is planted with its full planting density and it get its required water and fertilizer. Whereas, wheat or faba bean considered as the secondry crop, which it use the applied water and fertilizer for the main crop (sugar beet). To reduce intraspecific competition between the main crop (sugar beet) and the secondary crop (wheat or faba bean), the optimum planting density for either crops is 25%.

Another important crop in intercropping systems is tomato either winter or summer crop. In the winter, either wheat (Abd El-Zaher et al. 2013) or faba bean can intercrop with tomato (Ibrahim et al. 2010) systems. These two systems modified the microclimate for tomato, where wheat or faba bean plants protect tomato plants from low temperature in January and February. Fernandez-Munoz et al. (1995) indicated that exposing tomato plants to low temperature reduces pollen production, shed, viability, and tube growth. Furthermore, intercropping wheat with tomato can increase water use efficiency, where tomato grows tap deep strong root systems, which facilitate the absorption of soil moisture deeper than wheat root systems. In addition, the roots of tomato leave the soil in a good mechanical condition and it contains more fertilizing elements than those of most other crops (Pressman et al. 1997).

Regarding to summer tomato, two important intercropping systems were implemented, namely maize intercropped with tomato (Mohamed et al. 2013) and sunflower intercropped with tomato (Kestha and El-Baz 2004). Both crops can modify the microclimate for tomato, where it protected the tomato fruits from sun damage in July and August (Mohamed et al. 2013). Abdelmageed et al. (2003) stated that the reproductive processes in tomato are more sensitive to high temperatures than the vegetative processes. Maize intercropped with tomato proved to reduced pests and diseases that usually exist in tomato monoculture (Pino et al. 1994). Hao (2013) indicated that maize intercropping with tomato has control effect on powdery mildew that occurs in tomato plants. Furthermore, the spatial arrangment and defferent pattern of roots (shallow versus deep) exploit soil nutrients in the system and minimize plants competition (Ijoyah and Fanen 2012).

With respect to sunflower intercropped with tomato, Abdel (2006) indicated that intercropping sunflower with tomato resulted in mitigating heat stress through evaporative cooling and shading means, which improved fruit set and consequently yield quality.

Sugarcane offers another unique potential for intercropping. There are two growth cycles for sugarcane in Egypt: spring and fall sugarcane. Spring sugarcane is cultivated in February and harvested in February in the following year, where its growing season is 12 months. Fall sugarcane is cultivated in September and harvest 16 months later. The crop is planted in wide rows (100 cm), and takes several months to develop its canopy, during that time the soil and solar energy goes to waste. The growth rate of sugarcane during its early growth stages is slow, with leaf canopy providing sufficient uncovered area for growing of another crop (Nazir et al. 2002). In this case, the intercropped crop will not need any extra irrigation water as it will use the applied water to sugarcane to fulfill its required water. Furthermore, intercropping on sugarcane provides extra income for farmers during the early growth stage of sugarcane.

Regarding to spring sugarcane, sunflower can be intercropping with it. El-Gergawi et al. (2000) indicated that land productivity was increased when sunflower was intercropped with spring sugarcane. However, Abou-Keriasha et al. (1997) indicated that competition over solar radiation between sunflower plants and sugarcane plants was high because sunflower plants were longer than sugarcane plants in that growth stage. On the other hand, intercropped sesame with spring sugarcane did not cause competition over solar radiation between sesame plants and sugarcane plants because sesame leaves are erect and do not cause any shading over the growing sugarcane plants (Abou-Keriasha et al. 1997).

Furthermore, wheat intercropped with fall sugarcane (Ahmed et al. 2013) and faba bean intercropped with fall sugarcane are two common intercropping systems in Egypt. It provides extra income for farmers during the early growth stage of sugarcane. In addition, intercropping faba bean with sugarcane reduces the applied nitrogen fertilizer to sugarcane in this growth stage (Farghly 1997).

Fruit trees also provide good environment for intercropping, especially young evergreen fruit tree (1–3 years old) or deciduous fruit trees. Both wheat and faba bean can be planted under it in the winter.

Other common intercropping systems in Egypt are relay intercropping cotton on wheat which increase wheat cultivated area by the area assigned to be cultivated by cotton and can save the first and the second irrigations applied for cotton (Zohry 2005).

Intercropping Effect on Wheat Cultivated Area

Several intercropping systems can be used to increase wheat national production as follows:

- Wheat intercropped with tomato, where we assumed that 45% of winter tomato cultivated area will be assigned to intercropping with wheat. Wheat productivity under this system will be 70% of its productivity under sole planting.
- Wheat intercropped with sugar beet, where we assumed that 25% of early sugar beet (planted in October) cultivated area will be assigned to intercropping with wheat. Wheat productivity under this system will be 50% of its productivity under sole planting.
- Relay intercropping cotton on wheat, where we assumed that 90% cotton cultivated area will be assigned to intercropping with wheat. Wheat productivity under this system will be 80% of its productivity under sole planting.

The above three intercropping systems should be implemented on raised beds in the old lands.

- Wheat intercropped on 16% of sugarcane cultivated area and 16% of the cultivated area of fruit trees. Wheat productivity under this system will be 60% of its productivity under sole planting.

Table 4.1 indicated that the above intercropping systems can cause an increase in the cultivated area of wheat by 270,440 heaters, which can result in an increase in wheat total production by 1173,129 tons. The third agro-climatic zone contributed highly in this increase through the cultivated area of tomato. The second agro-climatic zone contributed highly in this increase through the cultivated areas of

Table 4.1 New added area and total production (ton) of wheat in old and new lands under intercropping systems in the agro-climatic zones of Egypt

Zone	Added area (ha)						Average productivity	Total production
	Area	Tomato	Sugar beet	Cotton	Sugarcane	Fruits		
1	Old	963	14,651	27,908	0	938	4.59	203,928
	New	140	670	335	0	283	4.25	6076
2	Old	63	12,495	33,585	0	9074	3.77	208,222
	New	28	7528	5862	0	34,357	4.49	214,578
3	Old	7905	7699	15,971	0	17,051	4.90	238,077
	New	3557	3009	3924	0	14,222	3.45	85,239
4	Old	2107	4870	2730	3321	5657	4.84	90,513
	New	948	1096	0	119	2238	4.37	19,217
5	Old	489	0	0	15,759	2772	4.41	83,898
	New	2543	0	0	2347	1229	3.82	23,381
Total		18,743	52,018	90,314	21,545	87,820		1173,129
Grand total area	270,440							

sugar beet and cotton. Furthermore, the fifth agro-climatic added the largest amount of wheat yield through the cultivated area of sugarcane (Table 4.1).

Intercropping Effect on Faba Bean Cultivated Area

With respect to the suggested intercropping systems for faba bean, there are four successful systems implemented in Egypt as follows:

– Faba bean intercropped with tomato, where we assumed that 45% of winter tomato cultivated area will be assigned to intercropping with faba bean. Faba bean productivity under this system will be 80% of its productivity under sole planting.
– Faba bean intercropped with sugar beet, where we assumed that 25% of early sugar beet cultivated area will be assigned to intercropping with wheat. Faba bean productivity under this system will be 50% of its productivity under sole planting.
– Faba bean intercropped on 16% of sugarcane cultivated area and 16% of the cultivated area of fruit trees. Faba bean productivity under this system will be 60% of its productivity under sole planting.

Similar to wheat, the faba bean intercropping systems can increase the cultivated area of faba bean by 210,306 ha, which can result in an increase in faba bean total production by 341,437 tons (Table 4.2). Large contribution of this area comes from the second agro-climatic zone, where sugar beet and fruit trees have a part of it. Tomato cultivated area also has a contribution in the third agro-climatic zone (Table 4.2).

Table 4.2 New added area and total production (ton) of faba bean in old and new lands under intercropping systems in the agro-climatic zones of Egypt

Zone	Added area (ha)						Average productivity	Total production
	Area	Tomato	Sugar beet	Sugarcane	Fruits	Total		
1	Old	1070	16,279	0	938	18,286	2.30	42,107
	New	156	745	0	283	1184	1.85	2185
2	Old	1639	13,884	0	9074	24,597	1.81	44,398
	New	7347	8364	0	34,357	50,068	1.02	51,274
3	Old	8783	8555	0	17,051	34,388	1.72	59,222
	New	5979	3343	0	14,222	23,545	1.30	30,494
4	Old	2341	5411	3321	5657	16,730	2.06	34,462
	New	8806	4871	119	2238	16,034	1.89	30,265
5	Old	543	0	15,759	2772	19,073	1.93	36,725
	New	2825	0	2347	1229	6401	1.61	10,306
Total		39,490	61,451	21,545	87,820	210,306		341,437

Intercropping Effect on Maize Cultivated Area

There are four well-known intercropping systems, where maize is the secondary crop as follows:

- Maize intercropped with soybean on 50% of its cultivated area and it produced 75% of its productivity under sole planting.
- Maize intercropped with summer tomato on 45% of its cultivated area and it produced 60% of its productivity under sole planting.
- Maize intercropped with peanut on 30% of its cultivated area and it produced 30% of its productivity under sole planting.
- Maize intercropped with sorghum on 33% of its cultivated area and it produced 50% of its productivity under sole planting.

Implementing the suggested intercropping systems for maize can increase its total area by 204,351 ha and maize total production will increase by 907,049 tons (Table 4.3). Of this area, the highest contribution of soybean existed in the fourth agro-climatic zone. The second agro-climatic zone contained the highest contribution of summer tomato. The third and the fourth agro-climatic zones contained the highest cultivated area of peanut and sorghum, respectively (Table 4.3).

Intercropping Effect on Sunflower Cultivated Area

Similar to maize, there are four well-known intercropping systems, where sunflower is the secondary crop as follows:

Table 4.3 New added area and total production (ton) of maize in old and new lands under intercropping systems in the agro-climatic zones of Egypt

Zone	Added area (ha)						Average productivity	Total Production
	Area	Soybean	Tomato	Peanut	Sorghum	Total		
1	Old	215	4284	0	0	4499	5.09	22,896
	New	0	4858	0	0	4858	4.56	22,164
2	Old	451	4890	1047	0	6388	4.47	28,539
	New	198	12,766	355	0	13,318	3.92	52,152
3	Old	63	5148	12,728	49,299	67,238	4.88	327,892
	New	0	3724	3518	463	7704	2.72	20,946
4	Old	6352	2671	3013	73,600	85,636	4.53	387,508
	New	12	1679	956	2550	5198	3.66	19,038
5	Old	0	276	6	4781	5064	3.07	15,523
	New	0	125	1280	3044	4448	2.34	10,391
Total		7291	40,421	22,902	133,737	204,351		907,049

- Sunflower intercropped with soybean on 20% of its cultivated area and it produced 20% of its productivity under sole planting.
- Sunflower intercropped with summer tomato on 45% of its cultivated area and it produced 60% of its productivity under sole planting.
- Sunflower intercropped with sugarcane and under fruit trees on 16% of its cultivated area and it produced 60% of its productivity under sole planting.

Under these suggested systems, sunflower total area can increase by 362,136 ha and sunflower total production can increase by 314,360 tons (Table 4.4). The highest contribution was found by soybean in the fourth agro-climatic zone. The second agro-climatic zone contained the highest contribution of summer tomato and fruit trees. The fifth agro-climatic zones contained the highest cultivated area of sugarcane (Table 4.4).

Intercropping Effect on Summer Forage Crops Cultivated Area

Intercropping can be also used to increase the availability of forage crops in the summer, represented by cowpea crop. Two intercropping systems can be implemented as follows:

- Intercropping cowpea with maize, where it can be implemented on 30% of its cultivated area and cowpea produce 60% of its sole planting.
- Intercropping cowpea with sunflower, where it can be implemented on 70% of its cultivated area and cowpea produce 60% of its sole planting.

The area of cowpea can be increased by 661,439 ha, which can produce 13,715,532 green yield of cowpea (Table 4.5). Because maize cultivated area is large, it

Table 4.4 New added area and total production (ton) of sunflower in old and new lands under intercropping systems in the agro-climatic zones of Egypt

Zone	Added area (ha)						Average productivity	Total production
	Area	Soybean	Tomato	Sugarcane	Fruits	Total		
1	Old	86	4284	0	685	5055	1.17	5927
	New	0	4858	0	777	5635	0.86	4822
2	Old	451	10,867	0	56,715	68,033	0.50	33,780
	New	396	28,368	0	214,730	243,494	0.91	221,303
3	Old	25	5148	0	824	5997	1.58	9454
	New	0	3724	0	596	4320	0.51	2222
4	Old	2541	2671	3321	427	8960	2.33	20,892
	New	5	1679	119	269	2071	0.87	1811
5	Old	0	276	15,759	44	16,079	0.80	12,863
	New	0	125	2347	20	2492	0.52	1286
Total		3504	62,001	21,545	275,087	362,136		314,360

Table 4.5 New added area and total production (ton) of cowpea in old and new lands under intercropping systems in the agro-climatic zones of Egypt

Zone	Added area (ha)				Average productivity	Total production
	Area	Maize	Sunflower	Total		
1	Old	21,461	22	21,483	21.0	451,143
	New	3990	60	4050	19.5	78,977
2	Old	119,043	912	119,955	21.3	2555,046
	New	30,570	1824	32,395	19.8	641,412
3	Old	210,480	601	211,081	21.9	4622,666
	New	8982	438	9419	21.0	197,801
4	Old	229,230	687	229,917	19.5	4483,382
	New	9216	66	9281	21.3	197,688
5	Old	17,543	0	17,543	19.9	349,103
	New	6316	0	6316	21.9	138,315
Total				661,439		13,715,532

contributed highly in increasing cowpea cultivated area, especially the fourth agro-climatic zone. Whereas, the cultivated area of sunflower in the second agro-climatic zone contributed highly in increasing cowpea cultivated area (Table 4.5).

Table 4.6 showed that using intercropping systems for wheat, faba bean, maize, sunflower, and cowpea can increase its cultivated area to high percentage. Wheat and maize cultivated area can increase by 20 and 22%, respectively. Whereas faba bean, sunflower, and cowpea cultivated area can increase to very high percentage, reaching 611, 550 and 128,186%, respectively.

Table 4.6 Recorded and potential added cultivated areas of the several crops through using intercropping in the agro-climatic zones in Egypt

	Area (ha)	Wheat	Faba bean	Maize	Sunflower	Cowpea
Zone 1	Recorded	131,028	7597	36,358	118	0
	Recorded + added	176,915	27,067	45,715	10,808	25,533
Zone 2	Recorded	402,191	18,672	213,734	3909	0
	Recorded + added	505,182	93,337	233,440	315,436	152,350
Zone 3	Recorded	389,140	5105	313,517	1483	0
	Recorded + added	462,478	63,038	388,459	11,799	220,500
Zone 4	Recorded	343,042	2082	340,637	1075	0
	Recorded + added	366,127	34,846	431,470	12,107	239,198
Zone 5	Recorded	89,444	963	34,084	0	516
	Recorded + added	114,581	26,437	43,596	18,571	24,375
Total recorded (ha)		1,354,844	34,418	938,329	6,585	516
Total recorded + added (ha)		1,625,284	244,724	1,142,680	368,721	661,955
Percentage of increase (%)		20	611	22	5500	128,186

Cultivation of Three Crops Per Year

The common practice of cultivation in Egypt involves cultivation of two crops per year, namely winter crop followed by summer crop. Changing crop sequence from two crops per year (winter then summer crop) to three crops per year (winter, fall then summer crop) or (winter, early summer then late summer crop) can increase food and feed production and increase food security. Zohry and Ouda (2017a, b) discussed, thoroughly, examples of cultivation of three crops per year which are presented in Table 4.7. In this table, the prevailing crop sequence is the cultivation of winter crop (wheat, sugar beet, faba bean, garlic, or full season clover) then cultivation of summer crop (maize, rice, soybean or sunflower). They suggested cultivating short season clover early in the winter season before the regular winter crop to help in decreasing summer season feed gap. Furthermore, to reduce edible oil gap, they suggested cultivation of sunflower or soybean as early summer crop (Table 4.7).

We suggested cultivating 500,000 ha of short season clover as early winter crop and to cultivate 41,667 ha of sunflower as early winter crop. As a result of limited water resources, this will require additional irrigation water to be applied to this additional area, namely 2,712,413,528 m^3. Because implementing intercropping requires that the main crop in the system should be planted on raised beds, an amount can be saved and used to cultivate the third crop, namely short season clover or sunflower. The saved amount of water for each crop is presented in Table 4.8.

Table 4.7 Examples of cultivation of three crops per year

Prevailing crops sequence	WR (m³/ha)	Suggested crops sequence	WR (m³/ha)	Deviation* (m³/ha)
Wheat then maize	16,075	Clover (short season), wheat then maize	16,050	+25
Wheat then rice	18,800	Clover (short season), wheat then rice	18,137	+662
Sugar beet then rice	21,316	Clover (short season), sugar beet then rice	21,391	−75
Sugar beet then maize	18,566	Sugar beet, soybean then maize (late)	18,013	+553
Sugar beet then maize	18,566	Sugar beet, sunflower then maize (late)	17,999	+567
Faba been then maize	15,933	Faba bean, soybean then maize (late)	16,287	−354
Garlic then maize	16,400	Garlic, soybean then maize (late)	17,125	−725
Clover (full season) then maize	21,066	Clover (full season), sunflower then maize (late)	20,429	+554
Wheat then maize	16,075	Wheat, sunflower then maize (late)	16,238	−163

*(+) means saving in the applied water and (−) means increase in the applied water

Table 4.8 Saved irrigation water as a result of implementing intercropping systems on raised beds

	Saved irrigation water (m³)
Tomato (winter)	34,642,594
Sugar beet	319,814,218
Cotton	231,697,518
Maize	1,737,021,283
Soybean	28,884,331
Sunflower	5,653,053
Tomato (summer)	85,285,440
Peanut	43,592,858
Sorghum	234,822,233
Total	2,721,413,528

Suggested Cropping Pattern that Increases Food Security

The suggested cropping pattern presented in Table 4.9 resulted in an increase in the cultivated area of crops characterized by having a gap between production and consumption, such as wheat, faba been, maize, sunflower, and cowpea. Increase in the area of these crops can occur as a result of intercropping systems. Furthermore, these additional areas do not require the application of any extra irrigation water. In

Table 4.9 Suggested cropping pattern that increases food security in Egypt

	Cultivated area (ha)	Water requirements (m^3)
Winter crops		
Wheat	1,625,284	8,825,826,373
Faba bean	244,724	178,997,045
Clover	624,740	5,948,296,442
Short season clover	500,000	0
Onion	148,173	1,244,936,988
Tomato	70,173	347,661,962
Potato	102,285	339,729,364
Sugar beet	231,193	1,955,202,088
Others	680,192	2,858,335,063
Total	4,226,764	24,082,485,325
Summer crops		
Cotton	100,349	1,264,924,487
Rice	506,249	5,896,134,782
Maize	1,142,680	9,259,942,383
Soybean	14,130	147,263,643
Sunflower	350,150	48,883,220
Sunflower (early)	41,665	0
Potato	53,110	385,322,673
Tomato	89,825	824,943,808
Sugarcane	661,955	4,271,214,589
Fruit trees	134,656	9,789,068,856
Cowpea	510,776	0
Others	641,341	8,736,972,296
Total	4,246,886	40,953,837,404
Grand total	8,473,649	62,323,656,938

fact, some of the applied water to these intercropping systems will be saved as a result of changing cultivation method from basins or on narrow furrows to raised beds.

Furthermore, cultivation of short season clover or cultivation of sunflower as a middle crop to reduce the gap of summer feed or edible oil will require additional irrigation water, where it could be obtained from the save water from the implemented intercropping systems. Thus, it will not need extra water than what is assigned to agriculture. Table 4.9 indicated that the national cultivated area can be increased under the suggested cropping pattern to be 8,473,649 ha, with 35% increase, compared to current cultivated area in 2014/15. Moreover, the increase in the cultivated area will not consume extra irrigation water.

Conclusion

Food security in Egypt can be increased through polycropping, where intercropping techniques and cultivation of three crops per year can be implemented. Using intercropping systems for wheat, faba bean, maize, sunflower, and cowpea can increase its cultivated area to high percentages. Wheat and maize cultivated area can be increased by 20 and 22%, respectively. Whereas, faba bean, sunflower, and cowpea cultivated area can be increased to very high percentages, reaching 611, 5500 and 128,186%, respectively. Irrigation water needed to implement cultivation of a middle crop between winter and summer seasons, namely short season clover or sunflower can be obtained from cultivation of all the suggested cropping systems on raised beds. Thus, the proposed cropping pattern can increase the national cultivated area to be 8,473,649 ha, with 35% increase, compared to current cultivated area in 2014/15. Moreover, the increase in the cultivated area will not consume extra irrigation water.

References

Abd El-Zaher ShR, Shams AS, Mergheny MM (2013) Effect of intercropping pattern and nitrogen fertilization on intercropping wheat with tomato. Egypt J Appl Sci 28(9):474–489

Abd El-Zaher ShR, Gendy EK (2014) Effect of plant density and mineral and bio-nitrogen fertilization on intercropping faba bean with sugar beet. Egypt J Appl Sci 29(7):352–366

Abdel CG (2006) Improvement of tomato fruit-set under natural high temperature: 1-intercropping tomato with sunflower or corn. J Dohuk Univ 9(2):2–16

Abdelmageed AH, Gruda N, Geyer B (2003) Effect of high temperature and heat shock on tomato (Lycopersicon esculentum Mill.) Genotypes under controlled conditions. Conference on International Agricultural Research for Development Göttingen, 8–10 Oct 2003

Abou-Elela AM (2012) Effect of intercropping system and sowing dates of wheat intercropping with sugar beet. J Plant Product 3(12):3101–3116

Abouelenein R, Oweis T, Sherif M, Khalil FA, Abed El-Hafez SA, Karajeh F (2010) A new water saving and yield increase method for growing berseem on raised seed bed in Egypt. Egypt J Appl Sci 25(2A):26–41

Abou-Keriasha MA., Zohry AA, Farghly BS (1997) Effect of intercropping some field crops with sugar cane on yield and its components of plant cane and third ratoon. J Agric Sci Mansoura Univ 22 12:4163–4176

Ahmed AM, Ahmed Nagwa R, Khalil Soha RA (2013) Effect of intercropping wheat on productivity and quality of some promising sugarcane cultivars. Minia J Agric Res Develop 33 (4):557–583

Chowdhury MK, Rosario EL (1994) Comparison of nitrogen, phosphorous and potassium utilization efficiency in maize/mung bean intercropping. J Agric Sci 122(2):193–199

El-Gergawi ASS, Saif LM, Abou-Salama AM (2000) Evaluation of sunflower intercropping in spring planted sugarcane fields in Egypt. Assuit J Agric Sci 312:163–174

Eskandari H, Ghanbari A, Javanmard A (2009) Intercropping of cereals and legumes for forage production. Notulae Scientia Biologicae 1(1):7–13

Farghly BS (1997) Yield of sugar cane as affected by intercropping with faba bean. J Agric Sci Mansoura Univ 22(12):4177–4186

Fernandez-Munoz R, Gonzalez-Fernandez JJ, Cuartero J (1995) Variability of pollen tolerance to low temperatures in tomato and related with species. J Hortic Sci 70:41–49

Gallaher RN (2009) Management of agricultural forestry and fisheries enterprises. Vol. I: Multiple Cropping Systems Encyclopedia of Life Support Systems (EOLSS)

Gebru H (2015) A review on the comparative advantages of intercropping to mono-cropping system. J Biol Agric Healthc 5(9):215–219

Geno L, Geno B (2001) Polyculture production: principle, benefits and risk of multiple cropping. A report for the Rural Industry Research and Development Corporation (RIRDC), Publication, No. 01134

Gomez AA, Gomez KA (1983) Multiple cropping in the humid tropics of Asia. Ottawa 32p

Hao W (2013) Control effect of tomato and maize intercropping against tomato powdery mildew. Plant Dis Pests 4(2):22–24

Ibrahim S, Shaaban M, Gendy EK (2010) Intercropping faba bean with tomato. Egypt J Appl Sci 25(6A):167–181

Ijoyah MO, Fanen FT (2012) Effects of different cropping pattern on performance of maize-soybean mixture in Makurdi, Nigeria. Sci J Crop Sci 1(2):39–47

Kestha MM, El-Baz MG (2004) Studies on sunflower-tomato intercropping. In: Proceedings of 16th International Sunflower Conference, Fargo, ND USA, pp 514–518

Mazaheri D, Madani A, Oveysi M (2006) Assessing the land equivalent ratio (LER) of two corn (*Zea mays* L.) varieties intercropping at various nitrogen levels in Karaj, Iran. J Cent Eur Agric 7(2):359–364

Mohamed W, Ahmed NR, Abd El-Hakim WM (2013) Effect of intercropping dates of sowing and N fertilizers on growth and yield of maize and tomato. Egypt J Appl Sci 28(12B):625–644

Nazir MS, Jabbar A, Ahmad I, Nawaz S, Bhatti IH (2002) Production potential and economics of intercropping in autumn-planted sugarcane. Int J Agric Biol 41:140–141

Ouda S, Noreldin T (2017) Evapotranspiration data to determine agro-climatic zones in Egypt. J Water Land Dev 32(I–III):79–86

Pino M, De-Los A, Bertoh M, Espinosa R (1994) Maize as a protective crop for tomato in conditions of environmental stress. Cult Trop 15:60–63

Pressman E, Bar-Tal A, Shaked R, Rosenfeld K (1997) The development of tomato root system in relation to the carbohydrate status of the whole plant. Ann Bot 80:533–538

Sheha AM, Ahmed Nagwa R, Abou-Elela AM (2014) Effect of crop sequence and nitrogen levels on rice productivity. Ann Agric Sci 52(4):451–460

Szumigalski AR, Van-Acker RC (2008) Intercropping: land equivalent ratios, light interception, and water use in annual intercrops in the presence or absence of in-crop herbicides. Am Agron J 100:1145–1154

Tolera A (2003) Effects nitrogen, phosphorus farmyard manure and population of climbing bean on the performance of maize (*Zea mays* L.)/climbing bean (*Phaseolus vulgaris* L.) intercropping system in Alfisols of Bako. An M.Sc. Thesis Presented to the School of Graduate Studies of Alemaya University

Wu KX, Wu BZ (2014) Potential environmental benefits of intercropping annual with leguminous perennial crops in Chinese agriculture. Agr Ecosyst Environ 188:147–149

Zohry AA (2005) Effect of relaying cotton on some crops under bio-mineral N fertilization rates on yield and yield components. Ann Agric Sci 431:89–103

Zohry AA, Ouda S (2016a) Crops intensification to face climate induced water scarcity in Nile Delta region. In: Management of climate induced drought and water scarcity in Egypt: unconventional solutions. Springer Publishing House, Berlin

Zohry AA, Ouda S (2016b) Upper Egypt: management of high water consumption crops by intensification. In: Management of climate induced drought and water scarcity in Egypt: unconventional solutions. Springer Publishing House, Berlin

Chapter 5
Cropping Pattern to Face Water Scarcity

Samiha A.H. Ouda and Abd El-Hafeez Zohry

Abstract The objective of this chapter is to propose a cropping pattern that uses the amount of irrigation water assigned to agriculture more efficiently and increases food production in the five agro-climatic zones in Egypt. Our results indicated that two cropping patterns can be suggested to face water scarcity. The first cropping pattern will depend on changing crops cultivation method from flat or narrow furrows to raised beds, which will save 20% of the applied irrigation water. This amount will account for 10.0 billion cubic meters on the national level and can be used to cultivate new areas and increase the national cultivated area. The cultivated area under this suggested pattern can increase by 24%, compared to the area cultivated in 2014/15 growing season. The other suggested cropping pattern can be implemented through implementing cultivation on raised bed and polycropping, namely intercropping and cultivation of three crops per year. The cultivated area under this suggested pattern will increase by 53%, compared to the area cultivated in 2014/15 growing season. Both suggested cropping patterns can save on the applied irrigation and consequently increase food production.

Keywords Water stress · Polycropping · Intercropping techniques · Raised beds cultivation

Introduction

Water scarcity was defined as the lack of sufficient available water resources to meet the demands of water usage within a region (FAO 2012). There are many regions in the world already experiencing water scarcity situation. Water scarcity has a huge impact on food production. Without water people do not have a means of irrigating their crops and, therefore, to provide food for the fast growing population (Mark et al. 2002). According to the Rosegrant et al. (2006), agriculture, which accounts for about 70% of global water withdrawals, is constantly competing with domestic, industrial, and environmental uses for a scarce water supply. In

S.A.H Ouda et al., *Cropping Pattern Modification to Overcome
Abiotic Stresses*, SpringerBriefs in Water Science and Technology,
https://doi.org/10.1007/978-3-319-69880-9_5

attempts to fix this ever-growing problem, many have tried to form more effective methods of water management. One such method is irrigation management. Irrigation and its planning are demanding tasks, which involve a multidisciplinary approach to define and calculating all the relevant factors and parameters (Katerji and Rana 2008). Many of the irrigation systems do not use the water in the most efficient way. This requires more water than necessary to be used to ensure healthy crops (Rosegrant et al. 2006). Furthermore, improving water management is an important way to increase crop yields. By minimizing crop-water stress, the producer obtains more benefits from improved cultural practices and realizes the full yield of the cultivars available. But the soil seldom has just the right amount of water for maximum crop production; a deficiency or a surplus usually exists (Cooke 2012).

Egypt is one of these areas that is affected by water scarcity, although Egypt is gifted by the Nile River, which is the main source of water for irrigation and other uses. Egypt has reached a state, where the quantity of water available is imposing limits on its national economic development. Egypt will reach absolute water scarcity (500 m^3/capita/year), with population projections in 2025. More than 85% of the water withdrawal from the Nile is used for irrigated agriculture. Furthermore, surface irrigation is the major system in Egypt applied to 83% of the old cultivated land (Nile Delta and Valley). Most of the on-farm irrigation systems are low

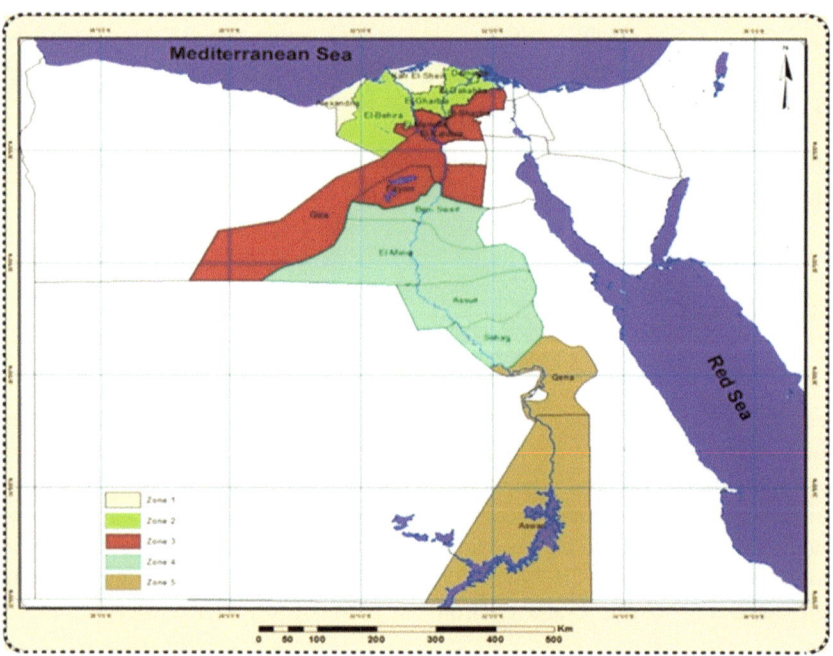

Fig. 1.1 Map of agro-climatic zones of Egypt using 10-year of ETo values

efficient coupled with poor irrigation management (Abou Zeid 2002). The agriculture strategy of Egypt for 2030 put emphasis on rationalization of irrigation water use, which is carried out using improved agricultural management practices on farm level to enhance water use efficiency and increase water and land productivity. This situation creates challenges for agricultural scientists to manage water properly, taking into consideration soil and water resources conservation.

The objective of this chapter was to propose a cropping pattern that uses the amount of irrigation water assigned to agriculture more efficiently and increases food production in the five agro-climatic zones in Egypt. Figure 1.1 showed the five agro-climatic zones developed by Ouda and Noreldin (2017).

Effect of Water Stress on Crops

To produce maximum yields, the soil must be able to provide water as it is needed by the crop. Crops are particularly sensitive to water stress when they are undergoing reproductive growth (Gardner et al. 1985). Thus, irrigation would help maintain optimal soil moisture during the growing period, thereby ensuring a more stable and higher agricultural production (Katerji and Rana 2008). On the other hand, irrigation is a method of transporting water to crops in order to maximize the amount of crops produced (Rosegrant et al. 2006).

Effect of Water Stress on Wheat

Exposing wheat plants to high moisture stress depressed seasonal consumptive use and grain yield (Bukhat 2005). During vegetative growth and under water stress, phyllochron decreases in wheat (McMaster 1997) and leaves become smaller, which could reduce leaf area index (Gupta et al. 2001) and number of reproductive tillers, in addition to limit their contribution to grain yield (Dencic et al. 2000). Furthermore, sensitivity for stress is the highest in stem elongation stage than in other growth stages. Drought stress during maturity resulted in about 10% decrease in yield, while moderate stress during the early vegetative period had essentially no effect on yield (Bauder 2001). Moreover, Zareian et al. (2014) reported that water stress during ear emergence and grain filling phases reduced grain yield and its components. Plants exposed to water stress in field have shown decreased CO_2 uptake and consequently decreased photosynthesis (Chaves et al. 2002). Mirzaei et al. (2011) reported that drought stress at all growth stages induced reducing grain yield and yield components. Drought stress at stages of stem elongation, flowering, and grain filling stages induced 32, 32 and 35% reduction in grain yield, respectively.

Effect of Water Stress on Maize

In maize, early season drought reduces plant growth and inhibits plant development (Heiniger 2001). Grzesiak (2001) reported that early stage of seedling growth and establishment is very sensitive to drought. Elongation of stem in maize under drought stress was reduced during vegetative stage (Ashraf 1989). Drought stress during vegetative growth stages especially during V1–V5, reduces growth rate, prolong vegetative growth stage, and conversely duration of reproductive growth stage is reduced (Aslam et al. 2013). Drought occurring between 2 weeks before and 2 weeks after silking stage can cause significant reductions in kernel set and kernel weight (Schussler and Westgate 1991), resulting in an average of 20–50% yield loss (Nielsen 2007). Andrade et al. (2000) indicated that water deficit stress decreases carbon availability and dry matter partitioning to ear at critical stages of ear development. Furthermore, pollination is most critical growth stage for water requirement (Aslam et al. 2013).

Effect of Water Stress on Faba Bean

Faba bean is more sensitive to drought than some other seed legumes including common bean, pea, and chickpea (Amede and Schubert 2003). Abd El-Mawgoud (2006) stated that the vegetative growth parameters, as well as yield components responded negatively to reduction in irrigation supply, namely plant height, number of leaves, and fresh and dry weights. Faba bean growth stages responded differently to drought. The phase of pod set was found to be more sensitive to drought, followed by pod filling and the vegetative phase (Younise 2002). Furthermore, in faba bean, water stress decreases the final leaf area (Saxena 1991), net photosynthesis (Hura et al. 2007), light use efficiency (Xia 1994), pod retention, and filling by reducing the availability of assimilate and distorting hormonal balances (Manschadi et al. 1998). However, mild water stress during flowering followed by sufficient irrigation of water after flowering resulted in slightly increased in seed yield and harvest index (Younise 2002).

Effect of Water Stress on Cotton

Worldwide cotton represents about 50% of the fiber used in textile industry. Cotton is also a complex plant. It is a perennial plant with an indeterminate habit, but it is cultivated as an annual crop (Jallas 1998). Water stress occurring during cotton growing season may reduce final lint yield. Numerous studies have reported that the effect of water stress on cotton yield depends on the timing and severity of the drought. Krieg (1997) indicated that the period from square initiation to first flower represents the most critical development period in terms of water supply affecting yield components. Water stress affects lint quality in numerous ways, especially during fiber elongation period. It causes a decrease in fiber length and fiber immaturity (Ritchie et al. 2004; McWilliams 2004; Mert 2005). Furthermore, the strength and elongation factors in cotton were well correlated with soil water (Johnson et al. 2002), where adequate soil water along with high ambient temperatures before and during boll development increased fiber maturity (Davidonis et al. 2004).

Effect of Water Stress on Rice

Water shortage is a very important factor on rice production, not only the amount of water deficit but also the time of deficit. Water deficit from booting to grain filling stage caused greatest rice yield reduction by 77%, than water stress during all the growth stages (vegetative, panicle initiation, and boot to grain fill) reduced grain yield and its components (Harbir and Ingram 2000). Water stress tends to delay flowering, and a larger delay in flowering was associated with a higher reduction in grain yield, harvest index, and percentages of fertile panicles and filled grains (Pantuwan et al. 2000). Drought for 2 weeks from 48 to 62 days after transplanting using withholding irrigation significantly reduced grain yield and grain weight (Ravindra et al. 2002). Rice growth period from booting to grain fill (reproductive stage) was the most sensitive to water deficit (Harbir et al. 2002).

Effect of Water Stress on Sugarcane

The knowledge of response of sugarcane to water stress under field condition is relatively limited. Most studies on water deficits in sugarcane have focused on irrigation management practices (Carr and Knox 2011). A number of physiological investigations conducted under laboratory or glasshouse conditions together with modeling studies have advanced the knowledge on water relations, stomatal functions, carbon assimilation, and dry matter partitioning in water-stressed sugarcane (Inman-Bamber and Smith 2005). There are four distinct growth stages in sugarcane, namely germination, tillering, grand growth, and maturity (Gascho and Shih 1983). The tillering and grand growth stages, known as the sugarcane formative phase, have been identified as the critical water demand period (Ramesh 2000). This is mainly because 70–80% of sugarcane yield is produced during this phase (Singh and Rao 1987). Ethan et al. (2016) indicated that water deficit during this stage significantly decreased sugarcane yield.

Suggested Cropping Pattern to Face Water Scarcity

1. Cultivation on raised beds

The main practice we suggest to face expected water scarcity is changing culti-vation method in the old lands from tradition method, which is done in basins or on narrow furrows to raised beds cultivation. Several papers showed the benefits of cultivating crops on raised beds nationally and internationally. Raised beds culti-vation proved to reduce the applied water to wheat by 20% and productivity in tons per fully irrigated hectare was increased by 15% (Abouelenein et al. 2009). Ahmad et al. (2009) reported that raised beds can save 20–25% of irrigation water, which increased water use efficiency. It also provided better opportunities to leach salts from the beds (Bakker et al. 2010). Hobbs et al. (2000) demonstrated that raised beds planting contributed significantly to improved water distribution and effi-ciency, increased fertilizer use efficiency and reduced weed infestation, lodging, and seed rate without sacrificing yield. Furthermore, Majeed et al. (2015) indicated that raised beds planting of wheat not only saves water but also improves fertilizer use efficiency and increases grain yield by 15%, compared to flat planting.

Other studies on raised beds cultivation showed that it reduces seed mortality rates and improves soil quality (Limon-Ortega et al. 2002), which led to enhanced root growth, and gave higher yield (Dey et al. 2015). Root length density was also more in upper 45 cm in beds due to porous soil environment (Dey et al. 2015).

Raised beds cultivation significantly and substantially increased maize growth, microbial functional groups, and enzyme activities compare to flat planting, thus it increases availability of essential crop nutrients by stimulating microbial activity (Zhang et al. 2012). Sing et al. (2010) found lower water consumption and higher wheat yield raised beds planting than under conventional flat beds planting due to decrease in irrigation amount. Beds planting also created better soil physical environment throughout the crop growth period, which led to higher crop pro-ductivity (Aggarwal and Goswami 2003). Research trials at India showed that raised beds were most suited for growing crops like maize, wheat, and soybean as they significantly decreased water use (Zhang et al. 2007).

Thus, we assumed that cultivation method for all crops cultivated in the current cropping pattern will be change to raised beds. As a result, 20% of the applied irrigation water will be saved. Furthermore, we assumed that sugarcane will be irrigated with gated pipes and fruit trees will be irrigated with drip system. The potential saved amounts of the irrigation water in each of the five agro-climatic zones in Egypt are presented in Table 1.1. The table indicated that the highest saved amount of irrigation water can come from the third agro-climatic zone, where the cultivated area in the old lands is the highest. The lowest saved amount can come from the first agro-climatic zone, where the cultivated area in the old lands is the lowest.

Table 1.1 Potential saved amounts of irrigation water (m^3) in the old lands in the agro-climatic zones in Egypt

	Zone 1	Zone 2	Zone 3	Zone 4	Zone 5
Winter crops					
Wheat	130,856,458	415,588,449	486,970,691	443,857,323	71,662,693
Faba bean	6,417,395	9,467,200	4,836,458	2,398,155	309,161
Clover	90,962,739	402,542,750	384,580,848	217,459,822	20,267,296
Onion	27,894,982	79,834,611	54,400,360	42,944,763	1,294,881
Tomato	2,232,441	3,747,970	21,314,382	5,899,950	1,447,850
Potato	2,505,164	28,127,713	21,327,000	7,104,000	2323
Sugar beet	110,265,035	100,150,316	65,817,145	43,581,721	0
Others	9,340,800	54,830,940	85,994,240	27,766,200	3,299,360
Total	380,475,016	1,094,289,950	1,125,241,124	791,011,934	98,283,565
Summer crops					
Cotton	74,938,819	98,514,900	49,094,500	9,149,299	0
Rice	227,088,488	673,874,689	215,952,556	0	0
Maize	54,674,028	328,219,821	598,364,808	701,880,939	53,881,688
Soybean	758,931	860,383	247,489	27,017,528	0
Sunflower	50,033	2,209,908	1,515,653	1,877,458	0
Potato	304,778	35,488,063	30,102,508	1,295,444	112,083
Tomato	19,388,218	23,943,807	25,891,590	14,541,771	1,520,054
Sugarcane	0	0	0	136,365,825	665,806,483
Fruit trees	22,561,000	236,877,910	466,758,350	166,867,733	84,476,058
Others	94,614,800	402,587,040	506,939,040	358,653,380	67,634,520
Total	494,379,094	1,802,576,521	1,894,866,494	1,417,649,377	873,430,887
Grand total	874,854,110	2,896,866,470	3,020,107,618	2,208,661,311	971,714,451

The above amount of irrigation water can be used to irrigate new area using sprinkler or drip systems depending on the cultivated crops. Table 1.2 presented the areas that can be cultivated using the saved irrigation water amounts presented in Table 1.1. The table showed that the highest cultivated area existed in the second agro-climatic zone, where the amount of saved water was high and water requirements for the cultivated crops in lower than the third agro-climatic zone. The lowest cultivated area existed in the fifth agro-climatic zone, as a result of high water requirements for the cultivated crops.

Table 1.3 showed the total potential saved amounts of water, the total potential cultivated area in Egypt as a result of using raised beds cultivation (added areas) and the total cultivated areas, which composed of old lands, new lands, and added lands. This practice will result in saving 9,972,203,960 m^3 which can be used to cultivate additional 1,184,168 hectares and increase the cultivated area to be 7,746,874 hectares (Table 1.3).

Table 1.2 Potential added cultivated area (ha) in the old lands as a result of cultivation on raised beds in the agro-climatic zones in Egypt

	Zone 1	Zone 2	Zone 3	Zone 4	Zone 5
Winter crops					
Wheat	27,036	80,959	89,298	77,237	11,917
Faba bean	1614	2267	1096	520	64
Clover	12,970	54,105	48,233	26,179	2307
Onion	3855	10,235	6540	4867	140
Tomato	606	929	4,977	1327	308
Potato	1070	11,606	8393	2720	1
Sugar beet	18,450	15,735	9696	6133	0
Others	1853	10,528	78,445	5008	576
Total	67,455	186,364	246,678	123,992	15,311
Summer crops					
Cotton	8786	10,573	5028	0	0
Rice	28,939	79,334	24,805	0	0
Maize	8687	48,184	85,194	92,784	7101
Soybean	0	128	0	3600	0
Sunflower	9	369	243	278	0
Potato	61	6688	5410	0	13
Tomato	2697	3079	3241	1682	174
Sugarcane	0	0	0	4843	22,981
Fruit trees	1660	16,069	30,194	10,017	4908
Others	6065	25,416	31,526	21,976	20,169
Total	56,905	189,841	185,641	135,179	55,346
Grand total	124,360	376,205	432,319	259,171	70,658

2. Cultivation on raised beds and implementing polycropping

Implementing cultivation on raised bed, with its saved water will increase the cultivated area. In addition, implementing intercropping systems and cultivation of three crops per year can increase the cultivated area with the same amount of irrigation water assigned to agriculture (see the details in Chap. 4). In Chap. 4, we calculated the national cultivated area to be 8,473,649 ha. This increase in the cultivated area can occur as a result of intercropping systems for wheat, faba bean, maize, sunflower, and cowpea, in addition to cultivation of short season clover and sunflower as middle crops between winter and summer seasons. These two middle crops will get its water requirements from the saved amount of irrigation water resulted from implementing intercropping system on raised beds. Thus, we added to it the area resulted from cultivation on raised bed, taking into consideration not to

Table 1.3 Potential total saved irrigation water, added cultivated area, and potential total area as a result of cultivation on raised beds

	Total saved water (m^3)	Added area (ha)	Total area (ha)
Winter crops			
Wheat	1,548,935,614	286,448	1,641,292
Faba bean	23,428,370	5,561	39,979
Clover	1,115,813,455	143,794	768,535
Onion	206,369,596	25,638	173,811
Tomato	34,642,594	8147	78,320
Potato	59,066,201	23,790	126,075
Sugar beet	319,814,218	50,014	281,207
Others	181,231,540	33,963	782,475
Total	3,489,301,587	577,354	3,891,692
Summer crops			
Cotton	231,697,518	24,387	124,735
Rice	1,116,915,732	133,079	639,328
Maize	1,737,021,283	241,949	1,180,278
Soybean	28,884,331	3727	17,857
Sunflower	5,653,053	899	7484
Potato	67,302,876	12,159	65,282
Tomato	85,285,440	10,873	100,698
Sugarcane	802,172,308	27,825	162,481
Fruit trees	977,541,051	62,848	611,722
Others	1,430,428,780	89,067	945,317
Total	6,482,902,373	606,814	3,855,182
Grand total	9,972,203,960	1,184,168	7,746,874

exceed the assigned amount of irrigation water to agriculture. Therefore, we subtracted the saved amount of irrigation water from cultivation of intercropping systems on raised beds, namely 2,721,413,528 m^3 from the total amount of the saved water from the intercropping and use the rest of the water to cultivate new lands using sprinkler or drip systems depending on crops type.

Table 1.4 presented the added area as a result of availability of irrigation water. The table showed that the saved water from crops that is the main crops in the suggested intercropping systems, namely winter tomato, sugar beet, maize, sunflower, summer tomato, sugarcane, and fruit trees will be assigned to cultivate short season clover and sunflower (middle crops). Therefore, there will be no added area from these crops. The added areas for the other crops will be resulted from intercropping and cultivation on raised beds.

Table 1.4 Potential added cultivated area (ha) in the old lands as result of cultivation on raised beds and polycropping in the agro-climatic zones in Egypt

	Zone 1	Zone 2	Zone 3	Zone 4	Zone 5
Winter crops					
Wheat	72,925	191,946	171,037	103,337	37,054
Faba bean	21,084	76,931	59,029	25,228	25,537
Clover	100,00	100,00	100,00	100,00	100,00
Short season clover	12,970	54,105	48,233	26,179	2307
Onion	3855	10,235	6540	4867	140
Tomato	0	0	0	0	0
Potato	1070	11,606	8393	2720	1
Sugar beet	0	0	0	0	0
Others	8226	10,528	78,445	5008	576
Total	120,131	355,352	371,678	167,339	65,615
Summer crops					
Cotton	0	0	0	0	0
Rice	28,939	79,334	24,805	0	0
Maize	9357	19,233	30,230	39,412	6577
Soybean	0	0	0	0	0
Sunflower	10,690	310,849	10,317	11,032	18,571
Sunflower (early)	8333	8333	8333	8333	8333
Potato	61	6688	5410	0	13
Tomato	0	0	0	0	0
Cowpea	25,533	152,350	220,500	239,198	23,859
Sugarcane	0	0	0	0	0
Fruit trees	0	0	0	0	0
Others	6065	25,416	185,641	21,976	20,169
Total	88,979	602,204	485,235	319,951	77,521
Grand total	209,110	957,556	856,912	487,291	143,136

Table 1.5 presented the total added area and the total national area after implementing cultivation on raised beds and polycropping. In this case, the added area will reach 3,154,006 hectares and the total cultivated area will reach 9,632,309 hectares.

Table 1.5 The added cultivated area and potential total area as a result of cultivation on raised beds and polycropping

	Added area (ha)	Total area (ha)
Winter crops		
Wheat	576,300	1,931,144
Faba bean	207,810	242,228
Clover	143,794	768,535
Short season clover	500,000	500,000
Onion	25,638	173,811
Tomato	0	70,173
Potato	23,790	126,075
Sugar beet	0	231,193
Others	102,783	782,475
Total	1,580,115	4,825,633
Summer crops		
Cotton	0	100,349
Rice	133,079	639,328
Maize	104,808	1,043,137
Soybean	0	14,130
Sunflower	361,459	368,043
Sunflower (early)	41,667	41,665
Potato	12,172	65,282
Tomato	0	89,825
Cowpea	661,439	661,955
Sugarcane	0	134,656
Fruit trees	0	548,874
Others	259,268	1,099,432
Total	1,573,891	4,806,676
Grand total	3,154,006	9,632,309

Conclusion

There are two suggested cropping patterns to face water scarcity. The first cropping pattern will depend on changing crops cultivation method from flat or narrow furrows to raised beds, which will save 20% of the applied irrigation water. This amount can be used to cultivate new areas and increase the national cultivated area. The cultivated area under this suggested pattern will increase by 24%, compared to the area cultivated in 2014/15 growing season. The other suggested cropping pattern can be implemented through implementing cultivation on raised bed and polycropping, namely intercropping and cultivation of three crops per year. The cultivated area under this suggested pattern will increase by 53%, compared to the area cultivated in 2014/15 growing season. Both suggested cropping patterns can save on the applied irrigation and consequently increase food production.

References

Abdel-Mawgoud AMR (2006) Growth, yield and quality of green bean (*Phaseolus vulgaris*) in response to irrigation and compost applications. J Appl Sci 2(7):443–450

Abou Zeid K (2002) Egypt and the world water goals. Egypt statement in the world summit for sustainable development and beyond, Johannesburg, South Africa

Abouelenein R, Oweis T, El Sherif M, Awad H, Foaad F, Abd El Hafez S, Hammam A, Karajeh F, Karo M, Linda A (2009) Improving wheat water productivity under different methods of irrigation management and nitrogen fertilizer rates. Egypt J Appl Sci 24(12A):417–431

Aggarwal P, Goswami B (2003) Bed planting system for increasing water use efficiency of Wheat (T. Aestibum) grown in Inseptisol. Indian J Agric Sci 73:422–425

Ahmad IM, Qubal B, Ahmad G, Shah NH (2009) Maize yield, plant tissue and residual soil N as affected by nitrogen management and tillage system. J Agric Biol Sci 1(1):19–29

Amede T, Schubert S (2003) Mechanisms of drought resistance in seed legumes. I. Osmotic adjustment. Ethiop J Sci 26:37–46

Andrade FH, Cirilo AG, Echarte L (2000) Factors affecting kernel number in maize. In: Physiological bases for maize improvement. (Otegui, ME, Slafer GA eds), Aslam M, Zamir MSI, Afzal I, Yaseen M, Mubeen M, Shoaib A). pp 59–74

Ashraf M (1989) Effect of water stress on maize cultivars during the vegetative stage. Artn Arid Zone 28:47–55

Aslam M, Zamir MSI, Afzal I, Yaseen M, Mubeen M, Shoaib A (2013) DOUGHT STRESS, ITS EFFECT ON MAIZE production and development of drought tolerance through potassium application. Cercetări Agronomice în Moldova. XLVI 2(154):347–351

Bakker D, Hamilton M, Hetherington GJ, Spann R (2010) Salinity dynamics and the potential for improvement of water logged and saline land in a Mediterranean climate using permanent raised beds. Soil Tillage Res 110(1):8–24

Bauder J (2001) Irrigating with limited water supplies. Montana State University Communications Services, Montana Hall, Bozeman, MT 59717, USA

Bukhat NM (2005) Studies in yield and yield associated traits of wheat *Triticum aestivum L.* genotypes under drought conditions. M.Sc. Thesis, Department of Agronomy, Sindh Agriculture University, Tandojam

Carr MKV, Knox JW (2011) The water relations and irrigation requirements of sugarcane (*Saccharum officinarum*): a review. Exp Agric 47:1–25

Chaves MM, Pereira JS, Maroco J, Rodrigues ML, Ricardo CPP, Osório ML, Carvalho I, Faria T, Pinheiro C (2002) How plants cope with water stress in the field? Photosynthesis and growth. Ann Bot. 89(7):907–916

Cooke R (2012) Water management. In: Illinois agronomy handbook. University of Illinois, USA, pp 143–152

Davidonis GH, Johnson AS, Landivar JA, Fernandez CJ (2004) Cotton fiber quality is related to boll location and planting date. Agron J 96:42–47

Dencic S, Kastori R, Kobiljski B, Duggan B (2000) Evaporation of' grain yield and its components in wheat cultivars and land races under near optimal and drought conditions. Euphytica 1:43–52

Dey D, Nath D, Jamatia PB (2015) Effect of raised bed planting method of maize under sandy loam soil of West Tripura. Intern J Appl Res 1(7):561–563

Ethan S, Olagoke O, Yunusa A (2016) Effect of deficit irrigation on growth and yield of sugarcane. Direct Res J Agric Food Sci 4(6):122–126

FAO (2012) FAOSTAT 2012: FAO Statistical database. FAO, Rome. http://faostat.fao.org/site

Gardner FP, Pearce RB, Mitchell RL (1985) Physiology of crop plants. Iowa State University Press, Ames

Gascho GJ, Shih SF (1983) Sugarcane. In: Teare ID, Peet MM (eds) Crop-water relations. Wiley, New York, pp 445–479

Grzesiak S (2001) Genotypic variation between maize (*Zea mays L.*) single cross hybrids in response to drought stress. Acta Physiologiae Plantarum 23:443–456

Gupta NK, Gupta S, Kumar A (2001) Effect of water stress on physiological attributes and their relationship with growth and yield in wheat cullivars at different growth stages. J Agron 86 (143):7–1439

Harbir S, Ingram KT (2000) Sensitivity of rice *Oryza sativa L.* to water deficit at three growth stages. Crop Res Hisar 20:355–359

Harbir S, Ingram KT, Jhorar RK (2002) Comparative performance of different water production functions for rice *Oryza sativa L.* Crop Res Hisar 23:203–213

Heiniger RW (2001) The impact of early drought on corn yield. North Carolina State University. http://www.ces.ncsu.edu/plymouth/cropsci/docs/early_drought_impact_on_corn.html

Hobbs PR, Singh Y, Giri GS, Lauren JG, Duxbury JM (2000) Direct seeding and reduced tillage options in the rice-wheat systems of the Indo-Gangetic plains of South Asia. IRRI workshop, Bangkok, pp 25–26

Hura T, Hura K, Grzesiak M, Rzepka A (2007) Effect of long term drought stress on leaf gas exchange and fluorescence parameters in C3 and C4 plants. Acta Physiol Plant 29:103–113

Inman-Bamber NG, Smith DM (2005) Water relations in sugarcane and response to water deficits. Field Crops Res 92:185–202

Jallas E (1998) Improved model-based decision support by modeling cotton variability and using evolutionary algorithms. Ph.D. dissertation. Mississippi State University, Mississippi State, (Diss. Abstr. 9829786)

Johnson RM, Downer RG, Bradow JM, Bauer PJ, Sadler EJ (2002) Variability in cotton fiber yield, fiber quality, and soil properties in a southeastern coastal plain. Agron J 94:1305–1316

Katerji N, Rana G (2008) Crop evapotranspiration measurement and estimation in the Mediterranean region, ISBN 978 8 89015 2412. INRA-CRA, Bari

Krieg DR (1997) Genetic and environmental factors affecting productivity of cotton. In: Dugger P, Richter DA. (eds), Proceeding Beltwide Cotton Conference, New Orleans. National Cotton Council America, Memphis, 7–10 Jan, p 1347

Limon-Ortega A, Sayre KD, Drijber RA, Francis CA (2002) Soil attributes in a furrow-irrigated bed planting system in northwest Mexico. Soil Till Res 63:123–132

Majeed A, Muhmood A, Niaz A, Javid S, Ahmad ZA, Shah SSH, Shah AH (2015) Bed planting of wheat (*Triticum aestivum L.*) improves nitrogen use efficiency and grain yield compared to flat planting. Crop J 3:118–124

Manschadi AM, Sauerborn J, Stutzel H, Gobel W, Saxena MC (1998) Simulation of faba bean (*Vicia faba L.*) growth and development under Mediterranean conditions: model adaptation and evaluation. Eur J Agron 9:273–293

Mark W, Rosegrant XC, Cline SA (2002) World water and food to 2025: dealing with scarcity. The International Water Management Institute

McMaster GS (1997) Phonology, development, and growth of wheat (*Triticum aestivum L.*) shoot apex: a review. Adv Agron 59:63–118

McWilliams D (2004) Drought strategies for cotton. Cooperative Extension Service Circular 582 College of Agriculture and Home Economics. Available at: http://www.cahe.nmsu.edu/pubs/circulars. (Verified on 15 Oct 2007)

Mert M (2005) Irrigation of cotton cultivars improves seed cotton yield, yield components and fiber properties in the Hatay region, Turkey. Acta Agric Scand B 55:44–50

Mirzaei A, Naseri R, Soleimani R (2011) Response of different growth stages of wheat to moisture tension in a semiarid land. World Appl Sci J 12(1):83–89

Nielsen RL (2007) Assessing effects of drought on corn grain yield. Purdue University, West Lafayette. http://www.kingcorn.org/news/articles.07/Drought-0705.html

Ouda S, Noreldin T (2017) Evapotranspiration data to determine agro-climatic zones in Egypt. J Water Land Dev 32(I–III):79–86

Pantuwan G, Fukai S, Cooper M, Rajatasereekul S, O'Toole JC (2000) Field screening for drought resistance. ACIAR Proceedings No.101. Increased lowland rice production in the Mekong Region. In: Proceedings of an International Workshop held in Vientiane, Laos, 30 Oct–2 Nov 2001, pp 69–77

Ramesh P (2000) Effect of different levels of drought during the formative phase on growth parameters and its relationship with dry matter accumulation in sugarcane. J Agron Crop Sci 185:83–89

Ravindra K, Tedia K, Malaiya S, Yerne A (2002) Effect of drought on root and shoot growth, plant water status, canopy temperature and yield of rice. J Soils Crops 12:179–182

Ritchie GL, Bednarz CW, Jost PH, Brown SM (2004) Cotton growth and development. Cooperative Extension Service and the University of Georgia, College of Agricultural and Environmental Sciences. Bulletin 1252

Rosegrant MW, Ringler C, Benson T, Diao X, Resnick D, Thurlow J, Torero M, Orden D (2006) Agriculture and achieving the millennium development goals. The World Bank (Agriculture and Rural Development Department), Washington, DC

Saxena MC (1991) Status and scope for production of faba bean in Mediterranean countries. Options Méditerranéennes 10:15-20-47

Schussler JR, Westgate ME (1991) Maize kernel set at low water potential: II. Sensitivity to reduceassimilate supply at pollination. Crop Sci 31:1196–1203

Sing VK, Dwivedi BS, Shukla AK, Mishra RP (2010) Permanent raised bed planting of the pigeonpea-wheat system on a typic Ustochrept: effects on soil fertility, yield and water and nutrient use efficiencies. Field crops Res 116:127–139

Singh S, Rao PNG (1987) Varietal differences in growth characteristics in sugarcane. J Agric Sci 108:245–247

Xia MZ (1994) Effects of soil drought during the generative development phase of faba bean (Vicia faba L.) on photosynthetic characters and biomass production. J Agric Sci 122:67–72

Younise ID (2002) Effect of water stress at different growth stages on growth and productivity of faba bean (Vicia faba L.). M.Sc. Thesis, University of Khartoum

Zareian A, Abad HHS, Hamidi A (2014) Yield, yield components and some physiological traits of three wheat (Triticum aestivum L.) cultivars under drought stress and potassium foliar application treatments. Int J Biosci 4(5):168–175

Zhang J, Sun J, Duan A, Wang J, Shen X, Liu X (2007) Effects on different planting patterns on water use and yield performance of winter wheat in the Huang-Huai-Hai plain of China. Agric Water Manag 92:41–47

Zhang X, Ma L, Gilliam FS, Li QWC (2012) Effects of raised-bed planting for enhanced summer maize yield on rhizosphere soil microbial functional groups and enzyme activity in Henan Province. China. Field Crops Res 130:28–37

Chapter 6
Cropping Pattern to Face Salinity Stress

Abd El-Hafeez Zohry and Samiha A.H. Ouda

Abstract In Egypt, salt-affected soils exist in 830,000 hectares located in the first, second and third agro-climatic zones. This area is distributed by 36, 23 and 8% in these three agro-climatic zones, respectively Develop cropping pattern under the condition of salt-affected soils require implementation of crop rotations to minimize the harm effects of salinity on crops (Ouda et al. 2016). Thus, the objective of this chapter was to develop cropping pattern that can maximize crops production and combat salinity stress in the five agro-climatic zone of Egypt. Three crop rotations were developed to reduce the harmful effect of salinity on the crops grown in the first, second and third ago-climatic zones. Furthermore, cultivation on raised beds and polycropping was suggested as improved management practices to be implemented inside the suggested crop rotations. The use of these three practices was integrated with the prevailing cropping pattern and new cropping pattern was developed. The total cultivated area under the suggested cropping pattern that faces salinity stress was about 9.2 million hectares, with 46% increase in the cultivated area, compared to the implemented cropping pattern in 2014/15.

Keywords Salt-affected soils · Raised beds cultivation · Polycropping Intercropping techniques · Cultivation of three crops per year

Introduction

High salinity is a common abiotic stress factor that seriously affects crop production in some parts of the world, particularly in arid and semi-arid regions (Neumann 1995). The total global area of salt-affected soils has recently been estimated to be approximately 830 million hectares (Martinez-Beltran and Manzur 2005). A saline soil is generally defined as one in which the electrical conductivity (EC) of the saturation extract (EC_e) in the root zone exceeds 4 dS m^{-1} (approximately 40 mM NaCl) at 25 °C and has an exchangeable sodium of 15%. The yield of most crop plants is reduced at this EC_e, though many crops exhibit yield reduction at lower EC_e (Jamil et al. 2011).

© The Author(s) 2018
S.A.H Ouda et al., *Cropping Pattern Modification to Overcome Abiotic Stresses*, SpringerBriefs in Water Science and Technology,
https://doi.org/10.1007/978-3-319-69880-9_6

Salinity affects plants in different ways such as osmotic effects, specific-ion toxicity and/or nutritional disorders (Läuchli and Epstein 1990). The extent by which one mechanism affects the plant over the others depends upon many factors including the species, genotype, plant age, ionic strength and composition of the salinizing solution, as well as the organ in question (Jamil et al. 2011). Furthermore, salinity not only decreases the agricultural production of most crops, but also, effects soil physicochemical properties, and ecological balance of the area (Hu and Schmidhalter 2002). Saline environments affect plant growth in different ways, such as an accretion of ions to toxic levels, and a reduction of nutrient availability, which causes reduction in crops productivity (Läuchli and Epstein 1990). Furthermore, a reduction in water uptake by the plant is occurred (Rameeh and Gerami 2015). The used irrigation systems and type of irrigation water have contributed to a large extent in converting arable lands to saline lands (Ajmal Khan and Weber 2006).

Furthermore, sea level rise is considered one of the major reasons of salinity intrusion into soil and groundwater, which is the result of both natural and human-induced climate change (Khanom 2016). Considering other forms of lands, deltas are easy victims of sea level rise. For instance, several studies found that Mekong delta (Wassmann et al. 2004), Nile delta (Frihy 2003), and Ganges floodplain (Sarwar 2005) are facing constant inundation and saline intrusion due to exposure to the sea. A study from the World Bank in 2000 suggests that increased

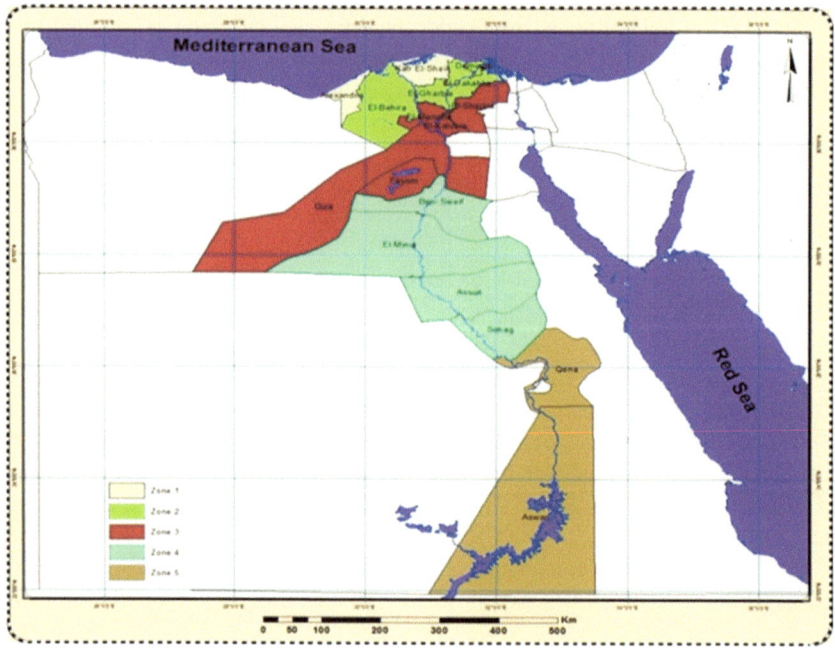

Fig. 6.1 Map of agro-climatic zones of Egypt using 10-year of ETo values

salinity alone will give rise to sea level by 0.3 m causing a net reduction of 0.5 million metric tons of rice production. Furthermore, the study predicted that sea level rise due to climate change will submerge a lot of low lying lands by 2050, and salinity intrusion will be more severe.

In Egypt, salt-affected soils exist in 830,000 hectares located in the first, second, and third agro-climatic zones. Developed cropping pattern under the condition of salt-affected soils require implementation of crop rotations to minimize the harmful effects of salinity on crops (Ouda et al. 2016). Thus, the objective of this chapter is to develop cropping pattern that can maximize crops production and combat salinity stress in the five agro-climatic zone of Egypt. Figure 6.1 showed the five agro-climatic zones developed by Ouda and Noreldin (2017).

Effect of Salinity on Crops Productivity

Using the salt-tolerant crops or salt-tolerant cultivars of a crop is the most important strategy to solve the problem of salinity. Salt tolerance in crops will also allow more effective use of poor quality irrigation water. To increase plant salt tolerance, there is a need for understanding the mechanisms of salt limitation on plant growth and the mechanism of salt tolerance at the whole-plant, organelle, and molecular levels (Munns et al. 2002). Many crops in Egypt are growing in salt-affected soil. The final yield of these crops is reduced in this type of soil, compared with its yield in the non-saline soil.

1. **Wheat crop**

Numerous studies have shown that wheat tiller appearance, abortion, or both are affected by salt stress (Nicolas et al. 1993). Wheat yield component affected most by salt stress is the number of spikes produced per plant (Maas and Grieve 1990). Salinity has inhibitory effects on wheat phenological aspects such as leaf number, leaf rate expansion, root/shoot ratio, and total dry matter yield (El-Hendawy et al. 2005). It often reduces shoot growth more than root growth (Läuchli and Epstein 1990). Salt stress, imposed while the shoot apex is in vegetative stage, can adversely affect spike development and decrease yields of wheat (Maas and Grieve 1990). When salinity was applied to wheat in seedling emergence, it had a profound influence on reproductive development (Grieve et al. 1993). Munns et al. (2006) found that wheat genotypes with the lowest Na^+ concentrations produced more dry matter than genotypes with high Na^+ concentrations. Nerson (1980) stated that soil salinity decreases grain yield of wheat more when plants are stressed prior to booting than when stressed later. Khan and Abdullah (2003) found that pollen viability in two wheat cultivars differing in salinity tolerance was reduced by 24–37%; depending on cultivar. It has long been known that salinity reduces the growth rate of the entire wheat plant and its specific organs, but it also affects plant development. The duration of plant development is also affected by salinity (Maas and Poss 1989).

2. **Sugar beet crop**

Sugar beet has the ability to grow in new soils that usually suffer from salinity and poor quality of irrigation water (Abdel-Mawly and Zanouny 2004). Rhoades et al. (1980) indicated that sugar yield of sugar beet was not affected by salinity level of 7.0 dS/m. Seed germination and seedling root length were significantly affected by the irrigation water with EC up to 8 and 4 dS/m, respectively (Jafarzadeh and Aliasgharzad 2007). Salinity retards sugar beet plant growth through its influence on the Osmatic adjustment and reducing nutrient uptake (Greenway and Munns 1980). Sugar beet plants grown under salinity stress showed imbalanced nutrient contents in their tissues. The effect of salt stress on the nutrient concentration in the plant varies among elements. Increasing salt concentration in growth media resulted in reducing K uptake by sugar beet plants (Shehata et al. 2000) and in turn, K content in shoots (Reda et al. 1980).

3. **Clover crop**

Egyptian clover is the major winter forage crop cultivated in the Nile Delta and Valley. It is the most widely grown field crop and occupies an area of 635,816 hectares in 2014/15 growing season. Egyptian clover is playing a vital role in the sustainability of Egyptian agriculture. It nourished the soils, suppressing weeds and providing a disease break in cereal-dominated crop rotations. It can be mowed several times for forage and then plowed under as green manure, which helps to increase the organic matter content of the soil (El-Nahrawy 2008). Egyptian clover is an important nitrogen-fixing legume crop and yet, molecular breeding programs for salt tolerance in this crop are rare. This could be largely due to the lack of detailed understanding of the responses of Egyptian clover to salt stress (Hussein et al. 2011). Aly et al. (2012) stated that salinity decreased slightly the stem of clover and leaves fresh weight. They also reported that the shoots and roots length of Egyptian clover decreased gradually with increasing NaCl level reaching to the minimum values at the highest dose of 100 mM NaCl. The responses of Egyptian clover plants to the highest level of salinity (100 mM NaCl) appeared also by decreasing in shoots and roots dry matter percentage compared with control.

4 **Rice Crop**

Rice is extremely sensitive to salinity during early vegetative growth (Pearson and Ayers 1966). Zeng and Shannon (2000) examined salinity effects on seedling growth and yield components of rice. They found that seedling growth was adversely affected at salinity levels as low as 1.9 dS/m, but this effect did not translate into a reduction in grain yield. They also found that seedling survival was adversely affected at salinities >3.4 dS/m. Salinity stress reduced number of florets per year, increase sterility and affect the time of flowering and maturity in rice (Khatun et al. 1995). Individual seed size was not significantly affected by salinity but grain yield per plant was reduced primarily by a reduction in the number of tillers per plant, number of spikelets per panicle, and the grain weight per panicle. Finally, they also found a substantial reduction in filled grains at 6 dS/m and higher suggesting that high salinity was causing some sterility. Salt stress in rice can reduce seedling emergence and when imposed at early vegetative stages, it reduces tillers and grain-bearing panicles leading to low yields (Pearson and Ayers 1966). However, certain rice cultivars can develop sterile spikelets, which appear to be genetically controlled, leading to further grain yield losses (Khatun et al. 1995).

5. **Cotton crop**

Cotton is one of the most salt-tolerant crops and was outstanding for successful cropping on salt-affected soils (Rhoades et al. 1980). However, the decline in cotton yield may be caused by the reduction of germination, delayed emergence and slow seedling growth (El-Saidi 1997). Cotton seedlings appeared particularly sensitive to soil salinity when B is present in an excessive amount in the soil solution (Mamani et al. 1998). In general, the most sensitive stage to salinity is flower bud formation, when growth can be completely arrested and high shedding induced. Cotton plants are much more resistant after flowering to salt concentration in the soil (El-Saidi 1997). However, cotton shows good compensation of plant number with plant size was found for soil EC up to 7.0 dS/m (Mamani et al. 1998). Irrigation of cotton with saline water containing up to 8000 ppm of soluble salts may produce acceptable yields in sandy soils, despite yield reductions with increasing salinity levels (Ahmad and Abdullah 1982). Cycles of irrigation with discontinuous use of saline water may help to prevent saline build up in the soil over time (Shennan et al. 1995).

Effect of Water and Salinity Stresses

Under field conditions, growing plants can suffer from both salinity (low soil osmotic potentials) and water deficit (low soil matric potentials) in the same time (Munns et al. 2002). However, some degree of both stresses can be occurring at different times and places in the root zone (Homaee et al. 2002). Clearly, the combination of stresses is more damaging than either one alone, but quantifying the growth-limiting contribution of each is difficult and can vary depending upon environmental conditions (Munns et al. 2002). Both stresses cause low water potentials in plants resulting in reduced leaf expansion rate, lower photosynthetic rate, and ultimately a reduced growth rate (Rawson and Munns 1984).

Management of Salinity

Salinization can be managed by changing farm management practices. We suggested cultivation on raised beds and implementation of crop rotations.

With respect to cultivation on raised beds cultivation in salt-affected soil, it gained importance for crops in many parts of the world (Sayre 2007). It proved to be a revolutionary means of preventing waterlogging and substantially increasing the productivity in Western Australia (Hamilton et al. 2000). Raised bed planting is a water-wise conservation agriculture-based practice, is gaining importance for high value-crops in irrigated agriculture (Devkota et al. 2015). It was reported that raised beds can save 20–25% of irrigation water, increasing water use efficiency (Ahmad et al. 2009) and providing better opportunities to leach salts from the furrows (Bakker et al. 2010). Favorable salt redistribution can be obtained with furrow irrigation that appeared effective for supporting cotton yields in saline conditions in Chile (Mamani et al. 1998).

Regarding crops rotation, it involves arrangement of crops planted on the same field; the succeeding crops should belong to different families (Huang et al. 2001). In salt-affected soil, appropriate crop rotation increases organic matter in the soil, improves soil structure, reduces soil degradation, and resulted in higher yields and greater farm profitability in the long term (Yazar et al. 2003). West and Post (2002) reported that use of crop rotation with legume crops reduce reliance on external inputs of nitrogen in salt-affected soils. Thus, the inclusion of clover in a rotation implemented in the salt-affected soil improves soil fertility more than the application of farmyard manure due to improvement in organic matter and physical conditions of the soil (Bhatti and Khan 2012). Abd El-Aal (1995) stated that the selection of crops during the reclamation of salt-affected soils should be based on tolerance to salt and waterlogging and its economic value. Annual crops with deep-rooted perennial species can prevent rising water tables bringing salts to the surface (Munns et al. 2002). Crop rotation improved soil structure (Raimbault and Vyn 1991), increased soil organic matter levels (Bremer et al. 2008), enhanced mycorrhizal associations (Johnson et al. 1995), increased water use efficiency (Tanaka et al. 2005), improved grain quality (Kaye et al. 2007), and reduced grain yield variability (Varvel 2000). Crop rotations also provide better weed control, interrupt insect and disease cycles, and improve crop nutrient use efficiency (Karlen et al. 1994). Furthermore, crop rotation can save on the applied irrigation water to crops. Said et al. (2016) indicated that a large amount of irrigation water can be saved when crops cultivated in the crop rotation planted on raised beds.

Suggested Crop Rotations

In Egypt, salinity exists in 830,000 hectares located in the first, second, and third agro-climatic zones. This area is distributed by 36, 23, and 8% in these three agro-climatic zones, respectively. Thus, we suggested implementing one crop rotation for soils with high salinity level, one crop rotation for soils with medium salinity level and another crop rotation for soils with low salinity level. Cultivation will be done on raised beds to save the applied irrigation water and use the saved amount to satisfy leaching requirements. The suggested rotations are three-year crop rotation, where the cultivated area is composed of three hectares. Each hectare is divided into three parts and each part is cultivated by winter and summer crops.

Crop Rotation Suitable for Low Salinity Level

The soil of this area contained low salinity level; therefore water requirements for the suggested crops in this rotation are not high. This rotation contained legume crops to maintain soil fertility, cereal crops, vegetable crops and fiber crop as followed:

First year (**section A**): Clover (Full season) then cowpea intercropped with maize.
First year (**section B**): Onion intercropped with sugar beet then tomato.
First year (**section C**): Relay intercropping cotton with wheat.
Second year: Section B, section C and section A.
Third year: Section C, section A and section B.

Crop Rotation Suitable for Medium Salinity Level

The soil of this area contained medium salinity level; therefore water requirements for the suggested crops in this rotation are high and contain rice (33.3%) and full season clover (33.3%). Both crops have high water requirements help in salt leaching from the soil. This rotation contained legume crops to maintain soil fertility, cereal crops, sugar crop, and fiber crop as followed:

First year (**section A**): Clover (one cut) then cotton.
First year (**section B**): Wheat, rice then short season clover.
First year (**section C**): Faba bean intercropped with sugar beet then maize.
Second year: Section B, section C, and section A.
Third year: Section C, section A, and section B.

Crop Rotation Suitable for High Salinity Level

The soil of this area contained high salinity level; therefore water requirements for the suggested crops in this rotation are very high and contain rice cultivated twice (66.6%) and full season clover (33.3%) and short-season clover (16%). Both crops have high water requirements help in salt leaching from the soil. This rotation contained legume crops to maintain soil fertility, cereal crops, sugar crop and oil crop as followed:

First year (section A): Wheat, rice then short season clover.
First year (section B): Onion intercropped with sugar beet, sunflower then maize (late).
First year (section C): Clover (full season) then rice.
Second year: Section B, section C and section A.
Third year: Section C, section A, and section B.

Cropping Pattern that Faces Salinity Stress

The suggested cropping pattern to be implemented in salt-affected soil is the one presented in Chap. 5, which will involve cultivation on raised beds for all crops in the cropping pattern. Furthermore, polycropping will be implemented, namely intercropping systems, and three crops cultivation per year. However, the saved amount of irrigation water in salt-affected areas from changing cultivation method

Table 6.1 The potential total cultivated area (ha) as a result of cultivation on raised beds in the agro-climatic zones in Egypt under salinity

	Zone 1	Zone 2	Zone 3	Zone 4	Zone 5
Winter crops					
Wheat	148,331	464,529	471,294	420,279	101,361
Faba bean	8630	20,418	6113	2602	1027
Clover	66,806	282,461	245,966	137,067	15,262
Onion	20,692	64,897	45,092	36,578	2287
Tomato	2840	18,687	34,104	14,815	7044
Potato	5717	66,141	37,865	12,394	230
Sugar beet	79,902	101,107	56,511	32,650	0
Others	25,889	206,988	102,792	40,293	145,903
Total	358,807	1,225,227	999,737	696,679	273,114
Summer crops					
Cotton	37,004	51,970	26,731	3033	0
Rice	122,374	365,857	120,448	0	0
Maize	41,917	250,836	391,895	433,420	41,184
Soybean	430	946	125	16,327	0
Sunflower	124	4,193	1707	1353	0
Potato	3469	34,517	24,486	742	75
Tomato	22,041	41,606	22,697	11,349	1066
Sugarcane	0	0	0	26,341	136,139
Fruit trees	8693	283,818	223,234	59,357	29,911
Others	134,096	290,373	287,608	109,361	38,134
Total	370,148	1,324,115	1,098,931	661,284	246,509
Grand total	728,955	2,549,343	2,098,668	1,357,963	519,623

to raised beds will be used to leach salts from the cultivated area of the suggested crop rotations. Thus, the added areas, as a result of irrigation water availability, will be reduced and consequently, the total cultivated area will be reduced, in comparison with the cropping pattern presented in Chap. 5.

1. Cultivation on raised beds

We assumed that the saved amount of irrigation water as a result of cultivation on raised beds will be used as leaching requirements in the first, second and third agro-climatic zones. Table 6.1 presented the potential total cultivated areas in the suggested cropping pattern in the five agro-climatic zones under raised beds cultivation. The table showed that the total area of these three zones will be reduced as a result of reduction in the added area.

The total added area and the total area on a national level are presented in Table 6.2. Thus, in salt-affected soils the suggested cropping pattern will result in a reduction in the added cultivated area to be 1,084,826 hectares and reduction in the total cultivated area to be 7,504,005 hectares, in comparison with the cropping pattern presented in Chap. 5.

Table 6.2 The potential total added cultivated area and potential total area as a result of cultivation on raised beds under salinity conditions

	Added area (ha)	Total area (ha)
Winter crops		
Wheat	250,950	1,605,794
Faba bean	4370	38,789
Clover	122,822	747,562
Onion	21,373	169,546
Tomato	7317	77,490
Potato	20,064	122,349
Sugar beet	38,977	270,170
Others	91,127	771,319
Total	557,000	3,803,018
Summer crops		
Cotton	18,390	118,738
Rice	102,429	608,678
Maize	220,924	1,159,253
Soybean	3698	17,828
Sunflower	792	7376
Potato	10,179	63,289
Tomato	8935	98,759
Sugarcane	27,825	162,481
Fruit trees	56,139	605,013
Others	78,516	859,572
Total	527,826	3,700,987
Grand total	1,084,826	7,504,005

2. Cultivation on raised beds and implementing polycropping

Implementing cultivation on a raised bed, intercropping systems and cultivation of three crops per year can increase the cultivated area using the same amount of irrigation water assigned to agriculture (see the details in Chap. 4). In Chap. 4, we calculated the national cultivated area to 8,473,649 hectares. This increase in the cultivated area can occur as a result of intercropping systems for wheat, faba bean maize, sunflower, and cowpea, in addition to cultivation of short-season clover and sunflower as middle crops between winter and summer seasons. These two middle crops will get their water requirements from the saved amount of irrigation water resulted from the intercropping system, namely cultivation on raised beds. Thus, we added to it the area resulted from cultivation in a raised bed.

Table 6.3 presented the added area as a result of the availability of irrigation water. It worth noting that the saved water from crops that is the main crops in the

Table 6.3 The potential total cultivated area (ha) as result of cultivation on raised beds and polycropping in the agro-climatic zones in Egypt under salinity conditions

	Zone 1	Zone 2	Zone 3	Zone 4	Zone 5
Winter crops					
Wheat	177,699	549,989	546,494	446,379	126,498
Faba bean	21,090	77,909	59,412	27,310	26,500
Clover	66,806	282,461	245,966	137,067	15,262
Short season clover	64,000	77,000	92,000	100,000	100,000
Onion	20,692	64,897	45,092	36,578	2287
Tomato	2452	17,972	29,525	13,489	6736
Potato	5,717	66,141	37,865	12,394	230
Sugar beet	68,094	88,990	47,591	26,517	0
Others	92,608	206,988	285,526	40,293	145,401
Total	519,159	1,432,348	1,389,471	840,027	422,916
Summer crops					
Cotton	31,381	43,829	22,105	3033	0
Rice	122,374	365,857	120,448	0	0
Maize	42,346	228,543	341,328	380,049	40,661
Soybean	430	847	125	12,728	0
Sunflower	6960	243,263	10,974	12,107	18,571
Sunflower (early)	5333	6416	7666	8333	8333
Potato	3469	34,517	24,486	742	75
Tomato	20,315	39,235	19,715	9667	892
Cowpea	16,341	117,309	202,860	239,198	24,375
Sugarcane	0	0	0	21,498	113,158
Fruit trees	7630	271,445	195,456	49,340	25,003
Others	134,096	290,373	429,393	109,361	113,328
Total	390,676	1,641,635	1,374,557	846,057	344,394
Grand total	909,835	3,073,983	2,764,028	1,686,084	767,309

suggested intercropping systems, namely winter tomato, sugar beet, maize, sunflower, summer tomato, sugarcane, and fruit trees will be assigned to cultivate short season clover and sunflower (middle crops). Therefore, there will be no added area from these crops. Taking into consideration leaching requirements in the first, second, and third agro-climatic zone, the added areas for the other crops will result from intercropping and cultivation on raised beds.

Table 6.4 indicated that the total added area and the total area will be higher than the value presented in Table 6.2 as a result of implementing polycropping and it will be lower than its counterpart value presented in Chap. 5 as a result of using some of the saved water in leaching salts.

Table 6.4 Potential added cultivated area and potential total area as a result of cultivation on raised beds and polycropping under salinity conditions

	Added area (ha)	Total area (ha)
Winter crops		
Wheat	492,216	1,847,060
Faba bean	177,803	212,221
Clover	122,822	747,562
Short season clover	433,000	433,000
Onion	21,373	169,546
Tomato	0	70,173
Potato	20,064	122,349
Sugar beet	0	231,193
Others	29,595	770,817
Total	1,296,873	4,603,921
Summer crops		
Cotton	0	100,349
Rice	102,429	608,678
Maize	94,598	1,032,927
Soybean	0	14,130
Sunflower	285,289	291,874
Sunflower (early)	36,082	36,082
Potato	10,179	63,289
Tomato	0	89,825
Cowpea	599,567	600,083
Sugarcane	0	134,656
Fruit trees	0	548,874
Others	220,302	1,076,552
Total	1,348,446	4,597,318
Grand total	2,645,319	9,201,238

Conclusion

Three crop rotations were developed to reduce the harmful effect of salinity on the crops grown in the first, second and third ago-climatic zones. Furthermore, cultivation on raised beds and polycropping was suggested as improved management practices to be implemented inside the suggested crop rotations. The use of these three practices was integrated with the prevailing cropping pattern and new cropping pattern was developed. The total cultivated area under the suggested cropping pattern that faces salinity stress was 9,201,238 hectares, with 46% increase in the cultivated area, compared to the implemented cropping pattern in 2014/15.

References

Abd El-Aal AIN (1995) Macro and micro morphological studies on the water table affected layer in some soils of Egypt. Ph.D. thesis, Faculty of Agriculture Cairo University Giza, Egypt

Abdel-Mawly SE, Zanouny I (2004) Response of sugar beet (*Beta Vulgaris*, l.) to potassium application and irrigation with saline water. Ass Univ Bull Environ Res 7(1):37–41

Ahmad R, Abdulla Z (1982) Biomass production of food and fiber crops using highly saline water under desert conditions. In: Biosaline research; proceedings from the 2nd international workshop on biosaline research 1980. La Paz, pp 149–163

Ahmad IM, Qubal B, Ahmad G, Shah NH (2009) Maize yield, plant tissue and residual soil N as affected by nitrogen management and tillage system. J Agric Biol Sci 1(1):19–29

Aly AA, Khafaga AF, Omar GN (2012) Improvement the adverse effect of salt stress in Egyptian clover (*Trifolium Alexandrinum L.*) by AsA application through some biochemical and Rt-PCR markers. J Appl Phytotech Environ Sanit 1:91–102

Bakker D, Hamilton M, Hetherington GJ, Spann R (2010) Salinity dynamics and the potential for improvement of water logged and saline land in a Mediterranean climate using permanent raised beds. Soil Tillage Res 110(1):8–24

Bhatti UA, Khan MM (2012) Review soil management in mitigating the adverse effects of climate change. Soil Environ 3(1):1–10. Soil Science Society of Pakistan. (http://www.sss-pakistan.org)

Bremer E, Janzen HH, Ellert BH, McKenzie RH (2008) Soil organic carbon after twelve years of various crop rotations in an Aridic Boroll. Soil Sci Soc Am J 72:970–974

Devkota M, Gupta RK, Martius C, Lamers JPA, Sayre KD, Vlek PLG (2015) Soil salinity management on raised beds with different furrow irrigation modes in salt-affected lands. Agric Water Manag 152:243–250

El-Hendawy SE, Hua Y, Yakout GM, Awad AM, Hafiz SE, Schmidhalter U (2005) Evaluating salt tolerance of wheat genotypes using multiple parameters. Europ J Agron 22:243–253

El-Nahrawy MA (2008) Vital role of berseem in Egyptian agriculture. In: Ninth international conference on Dry land development, Alexandria

El-Saidi MT (1997) Salinity and its effect on growth, yield and some physiologicalprocesses of crop plants. In: Jaiwal PK, Singh RP, Gulati A (eds) Strategies for improving salt tolerance in higher plants. Science Publ. Enfield, New Hampshire

Frihy OE (2003) The Nile delta-Alexandria coast: vulnerability to sea-level rise, consequences and adaptation. Mitig Adapt Strat Glob Change 8(2):115–138

Greenway H, Munns R (1980) Mechanisms of salt tolerance in nonhalophytes. Ann Rev plant physiology 31:149–190

Grieve CM, Lesch SM, Maas EV, Francois LE (1993) Leaf and spikelet primordia initiation in salt-stressed wheat. Crop Sci 22:1286–1294

Hamilton G, Bakker D, Houlebrook D, Spann C (2000) Raised beds prevent waterlogging and increase productivity. J Dept Agric Western Australia, (41)1-27-33

Homaee M, Feddes RA, Dirksen C (2002) A macroscopic water extraction model for nonuniform transient salinity and water stress. Soil Sci Soc Am J 66:1764–1772

Hu Y, Schmidhalter U (2002) Limitation of salt stress to plant growth. In: Hock B, Elstner CF (eds) Plant toxicology. Marcel Dekker Inc., New York, pp 91–224

Huang Q, Yin Z, Tian C (2001) Effect of two different straw mulching methods on soil solute salt concentration. Arid Land Geogr 24:52–56

Hussein MM, Abd El-Kader AA, Alva A (2011) Response of growth, anatomical structure and mineral status of grain sorghum to foliar. J Crop Improv 24:324–336

Jafarzadeh A, Aliasgharzad N (2007) Salinity and salt composition effects on seed germination and root length of four sugar beet cultivars. Biologia 62(5):562–564

Jamil A, Riaz S, Ashraf M, Foolad MR (2011) Gene expression profiling of plants under salt stress. Crit Rev Plant Sci 30(5):435–458

Johnson MG, Levine ER, Kern JS (1995) Soil organic matter: distribution, genesis and management to reduce greenhouse gas emissions. Water Air Soil Pollut 82:593–615

Karlen DL, Varvel GE, Bullock DG, Cruse RM (1994) Crop rotations for the 21st century. Adv Agron 53:1–44

Kaye NM, Mason SC, Jackson DS, Galusha TD (2007) Crop rotation and soil amendments alters sorghum grain quality. Crop Sci 47:722–729

Khan MA, Abdullah Z (2003) Reproductive physiology of two wheat cultivars differing in salinity tolerance under dense saline-sodic soil. Food, Agric Envrion. 1:185–189

Khan M, Weber DJ (2006) Ecophysiology of high salinity tolerant plants. Springer, The Netherlands, pp 11–30

Khanom T (2016) Effect of salinity on food security in the context of interior coast of Bangladesh. Ocean Coast Manag 130:205–212

Khatun S, Rizzo CA, Flowers TJ (1995) Genotypic variation in the effect of salinity on fertility in rice. Plant Soil 173:239–250

Läuchli A, Epstein E (1990) Plant responses to saline and sodic conditions. In: Tanji KK (ed) Agricultural salinity assessment and management. ASCE manuals and reports on engineering practice No, 71. ASCE New York, pp 113–137

Maas EV, Grieve CM (1990) Spike and leaf development in salt-stressed wheat. Crop Sci 30:1309–1313

Maas EV, Poss JA (1989) Salt sensitivity of wheat at different growth stages. Irrig Sci 10:29–40

Mamani OR, Doussoulin EE, Serri GH (1998) Response of cotton (*Gossypiulm barbadense*) to nitrogen fertilizer, plant density and leaching water in the Lluta Valley. Idesia 15:49–58

Manchanda G, Garg N (2008) Salinity and its effects on the functional biology of legumes. Acta Physiol Plant 30:595–618

Martinez-Beltran J, Manzur CL (2005) Overview of salinity problems in the world and FAO strategies to address the problem. In: Proceedings international salinity forum, Riverside, CA

Munns R, Husain S, Rivelli AR, James RA, Condon AG, Lindsay MP, Lagudah ES, Schachtman DP, Hare RA (2002) Avenues for increasing salt tolerance of crops, and the role of physiologically based selection traits. Plant Soil 247(1):93–105

Munns R, Richard A, Läuchli A (2006) Approaches to increasing the salt tolerance of wheat and other cereals. J Exp Bot 57:1025–1043

Nerson H (1980) Effects of population density and number of ears on wheat yield and its components. Field drop Res 3:225–234

Neumann PM (1995) Inhibition of root growth by salinity stress: toxicity or an adaptive biophysical response.In: Baluska F, Ciamporova M, Gasparikova O, Barlow PW (eds) Structure and function of roots. Academic Publishers, The Netherlands, pp 299–304

Nicolas ME, Mums R, Samarakoon AB, Gifford RM (1993) Elevated CO_2 improves the growth of wheat under salinity. Aust J Plant Physiol 20:349–360

Ouda S, Noreldin T (2017) Evapotranspiration data to determine agro-climatic zones in Egypt. J Water Land Dev 32(I–III):79–86

Ouda SA, Zohry AEH, Khalifa H (2016) Combating deterioration in salt-affected soil in Egypt by crop rotations. In: Management of climate induced drought and water scarcity in Egypt: unconventional Solutions. Springer Publishing House, New York

Pearson GA, Ayers AD (1966) Relative salt tolerance of rice during germination and early seedling development. Soil Sci 102:151–156

Raimbault BA, Vyn TJ (1991) Crop rotation and tillage effects on corn growth and soil structural stability. Agron J 83:979–985

Rameeh V, Gerami M (2015) Soil salinity effects on phenological traits, plant height and seed yield in rapeseed genotypes. Soil Sci Annu 66(1):17–20

Rawson HM, Munns R (1984) Leaf expansion in sunflower as influenced by salinity and short-term changes in carbon fixation. Plant, Cell Environ 7:207–213

Reda KA, AA Shalaby, HT Kishk, Hegazi AM (1980) Some effects of potassium on growth yield and chemical composition of beet irrigated with saline water containing different levels of boron. Ain Shams Univ Fac Agric Res Bull 12337:16

Rhoades JD, Rawlins SL, Phene CJ (1980) Irrigation of cotton with saline drainage water. ASCE, Portland

Said AS, Zohry AA, Ouda S (2016) Unconventional solution to increase crop production under water scarcity. In: Major crops and water scarcity in Egypt. Springer Publishing House, New York, pp 99–114

Sarwar MGM (2005) Impacts of sea level rise on the coastal zone of Bangladesh. See. http://static.weadapt.org/placemarks/files/225/golam_sarwar.pdf

Sayre K (2007) Conservation agriculture for irrigated agriculture in Asia. In: Lal R, Suleimenov M, Stewart BA, Hansen DO, Doraiswamy P (eds) Climate change and terrestrial carbon sequestration in central Asia. Taylor and Francis, The Netherlands, pp 211–242

Shehata MM, Shohair AA, Mostafa SN (2000) The effect of soil moisture level on four sugar beet varieties. Egypt J Agric Res 78(3):1141–1160

Shennan C, Grattan SR, May DM, Hillhouse CJ, Schachtman DP, Wander M, Robets B, Afoya S, Burau RG, McNeish C, Zelinski L (1995) Feasibility of cyclic reuse of saline drainage in a tomato-cotton rotation. J Environ Qual 24(3):476–486

Tanaka DL, Anderson RL, Rao SC (2005) Crop sequencing to improve use of precipitation and synergize crop growth. Agron J 97:385–390

Varvel GE (2000) Crop rotation and nitrogen effects on normalized grain yields in a long-term study. Agron J 92:938–941

Wassmann R, Hien NX, Hoanh CT, Tuong TP (2004) Sea level rise affecting the Vietnamese Mekong Delta: water elevation in the flood season and implications for rice production. Clim Change 66(1e2), 89e107

West TO, Post (2002) Soil organic carbon sequestration rates by tillage and crop rotation: a global data analysis. Soil Sci Soc Am J 66:1930–1946

Yazar A, Hamdy A, Gencel B, Sezen MS, Kocc M (2003) Wheat yield response to irrigation by saline water under the Mediterranean climatic conditions in Turkey. Sustainable Use of Highly Saline Water for Irrigation of Crops under Arid and Semi-Arid Conditions: new Strategies. Çukurova University, Adana

Zeng L, Shannon MC (2000) Salinity effects on seedling growth and yield components of rice. Crop Sci 40:996–1003

Chapter 7
Future Water Requirements for Prevailing Cropping Pattern

Mostafa Morsy and Samiha A.H. Ouda

Abstract The objective of this chapter was to calculate the values of water consumptive use for the studied crops in the prevailing cropping pattern in 2030 under the effect of climate change in the five agro-climatic zones in Egypt. The projected values of soil temperature on the level of each agro-climatic zone in 2030 revealed that there will be an increasing trend in its value in 2030, compared to its counterpart values in 2015. The highest differences will occur in the summer season from May to September. Whereas, in the winter season, there will be small differences between soil temperature in 2015 and 2030 from September to December, then the difference becomes higher from January to April. Furthermore, soil moisture content in the layer of 0–10 cm in 2030 will be reduced to the degree that it will have the same value in the winter as the summer due to increase in soil evaporation, which will reduce soil moisture content to a very low level. These results were true for the five agro-climatic zones in Egypt. This, in turn, can result in increased water requirements for cultivated crops.

Keywords Soil temperature · Soil moisture content · Climate change · Water consumptive use

Introduction

Climate change may alter both water availability and crop water requirement significantly as a result of changing temperatures and precipitation. Water stress is likely to intensify in the Mediterranean countries (Portugal and Spain) and some parts of Central and Eastern Europe (Döll 2002). Substantial irrigation requirements may be expected in some countries where irrigation hardly exists at present (Holden and Brereton 2003). An increase in temperatures could also lead to a net deficit in atmospheric water content, thus excessive evaporation from soil, water, and plant surfaces would occur. Land ecosystems would require more water to match increased water demand, and consequently to prevent drought (Kimball et al. 2002). In some countries, like Turkey, a decrease in effective rainfall and consequently a

© The Author(s) 2018
S.A.H Ouda et al., *Cropping Pattern Modification to Overcome Abiotic Stresses*, SpringerBriefs in Water Science and Technology,
https://doi.org/10.1007/978-3-319-69880-9_7

decline in water resources are expected. Reference and crop evapotranspiration rates are likely to increase, as well as irrigation requirements for the cultivated crops (Onol et al. 2006).

IPCC (2013) stated that the new Representative Concentration Pathways (RCP2.6, RCP4.5, RCP6.0, and RCP8.5) produced by climate models from the recent report by IPCC in 2013 can attain more accurate assessment of the effect of climate change (Solomon et al. 2007). Although there is a level of uncertainty associated with the projections of climate change in the future using the developed models by IPCC, using these models is the only method available for the evaluation. To lower that uncertainty, Morsy (2015) developed a methodology to select the most suitable global climate model and scenarios in Egypt to assess the effect of climate change. This methodology depended on performing a comparison between measured meteorological data and the projected data from four global climate models represented by its four RCPs scenarios during the period from 2006 to 2012. His results indicated that RCP6.0 scenario from CCSM4 model was more accurate to project the meteorological variables because it achieved the closest values of the goodness of fit test between measured and projected meteorological data.

The Community Climate System Model (CCSM) is a coupled climate model for simulating the earth's climate system. The model is composed of four separate models simultaneously simulating the earth's atmosphere, ocean, land surface and

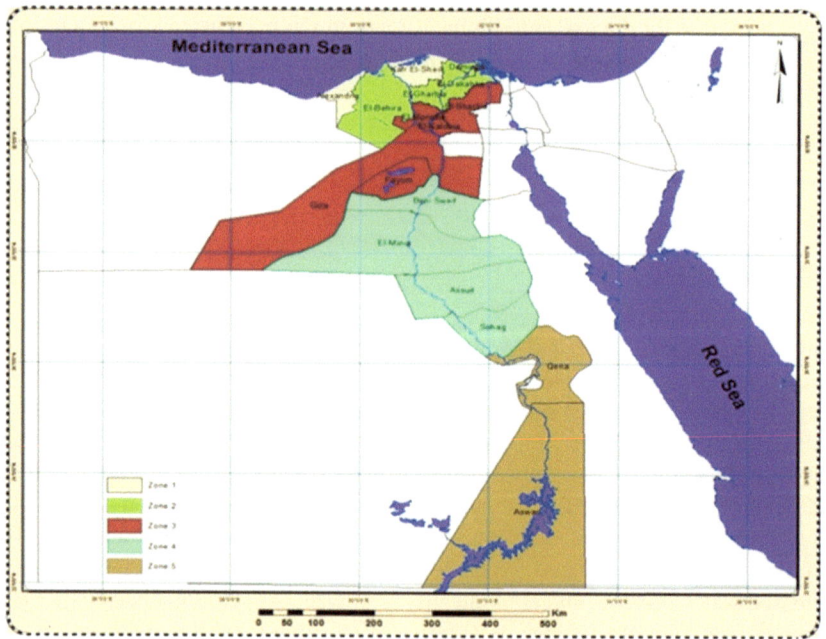

Fig. 7.1 Map of agro-climatic zones of Egypt using 10-year of ETo values

sea-ice, and one central coupler component. The CCSM allows researchers to conduct fundamental research into the earth's past, present, and future climate states. The model was developed by National Center for Atmospheric Research (NCAR), Community Climate System Model, USA. The model has a horizontal resolution equal to $1.25° \times 0.94°$ (http://www.cesm.ucar.edu/models/ccsm4.0/).

Furthermore, RCP6.0 is one of the four RCPs scenarios produced by CCSM4 model to represent a larger set of mitigation scenarios and have different targets in terms of radiative forcing in 2100. The scenario is a stabilization scenario, in which total radiative forcing is stabilized shortly after 2100, without overshoot, by the application of a range of technologies and strategies for reducing greenhouse gas emissions (Fujino et al. 2006; Hijioka et al. 2008).

The objective of this chapter was to calculate the values of water consumptive use for the studied crops in the prevailing cropping pattern in 2030 under the effect of climate change in the five agro-climatic zones in Egypt. Figure 7.1 showed the five agro-climatic zones developed by Ouda and Noreldin (2017).

Effect of Climate Change on Soil Temperature

Soil temperature, directly and indirectly, affects soil physical processes like infiltration and hydraulic conductivity (Ren et al. 2014). There is a close relationship between air temperature and soil temperature and a general increase in air temperature will inevitably lead to an increase in soil temperature. The temperature regime of the soil is governed by gains and losses of radiation at the surface, the process of evaporation, heat conduction through the soil profile, and convective transfer via the movement of gas and water (Karmakar et al. 2016). Temperature increase by climate change effect results in greater evapotranspiration loss of water from the soil. The rise in temperature increases the potential evapotranspiration and transpiration, if the plant canopy is not suffering from limited water supply due to climate or soil-induced drought, e.g., low precipitation, irrigation or limited water storage capacity (Rustad et al. 2000).

Soil Temperature in the Agro-Climatic Zones of Egypt

We collected data on soil temperature in the five agro-climatic zones in Egypt in 2015. The collected data was obtained from National Centers for Environmental Prediction; Climate Forecast System Reanalysis (NCEP/CFSR). Regular monthly means products with $0.5 \times 0.5°$ resolution for the soil layer between 0 and 10 cm. These values were compared to the value of RCP6.0 produced by CCSM4 climate model in 2030.

Figure 7.2a and b indicated that, in the first and second agro-climatic zone, soil temperature values in 2030 will be higher than its counterpart in 2015. The highest

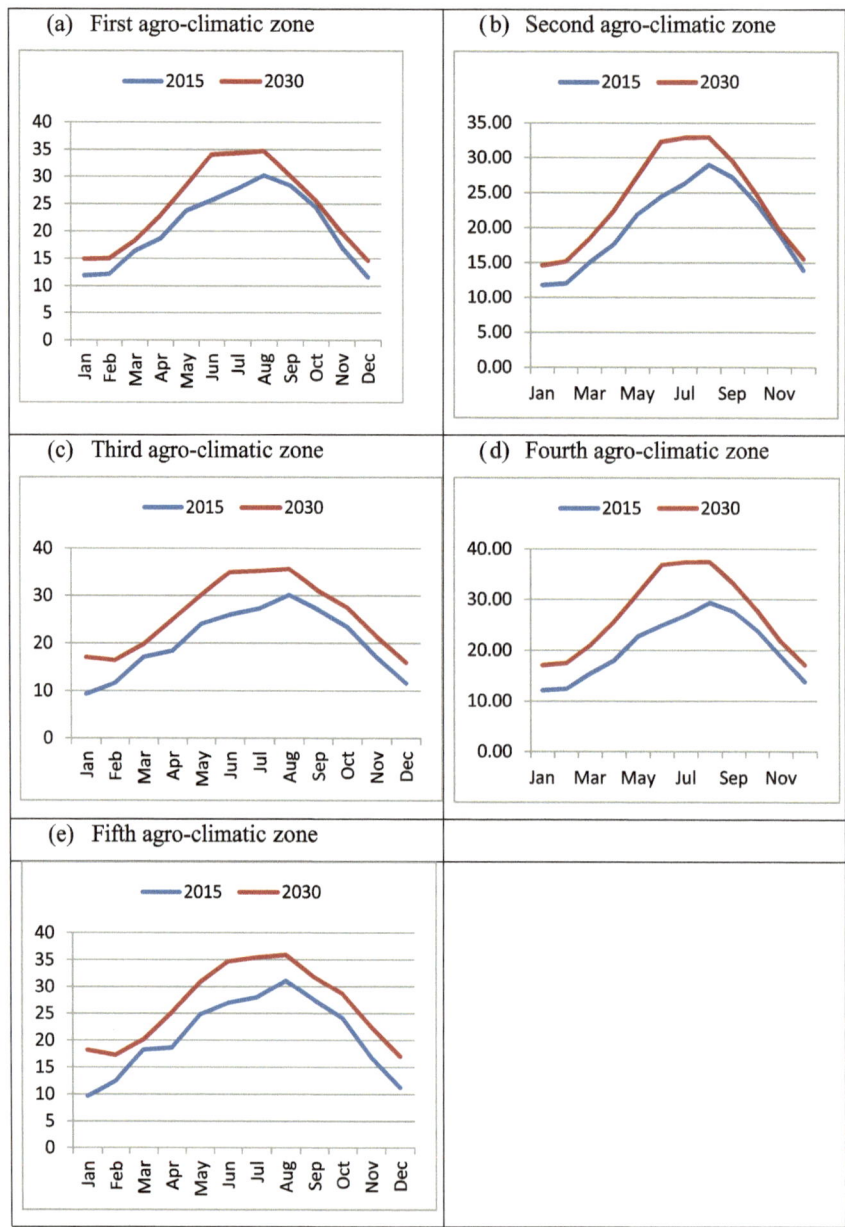

Fig. 7.2 Comparison between soil temperature in 2015 and 2030 in the agro-climatic zones of Egypt

differences will occur in the summer season from May to September. Furthermore, in the winter season, there will be small differences between soil temperature in 2015 and 2030 from September to December then the difference becomes higher from January to April.

Figure 7.2c indicated the same trend found in Fig. 7.2a and b; however, the difference between soil temperature values in 2015 and 2030 was higher. In Fig. 7.2d and e, it can be observed that the difference between soil temperature in 2015 and 2030 become higher during both winter and summer seasons.

The above results will have its implication on seed germination and evaporation from the soil in 2030, which will add another stress factor on the productivity of crops.

Effect of Climate Change on Soil Moisture

Although soil moisture constitutes only about 0.005% of global water resources, it is an important part of the water cycle and is a key variable controlling numerous processes and feedback loops within the climate system (Seneviratne et al. 2010). Soil moisture content is already being affected by rising temperatures and changes in precipitation amounts, both of which are evidence of changes in climate (Lopez-Moreno and Beniston 2009).

Water retention is a major hydrological property of soil because it governs soil functioning in ecosystems. While water-retention capacity is an intrinsic soil property based on clay content, structure and organic matter levels, soil moisture content is highly dynamic, and is, if based on natural factors only, the balance between rainfall, evapotranspiration, surface runoff, and deep percolation (Calanca et al. 2006). Maintaining water-retention capacity and porosity are important to reduce the impacts of intense rainfall and droughts, which are projected to become more frequent and severe in the future (García-Ruiz et al. 2011). Modeling of soil moisture content over the past 60 years in the Mediterranean region showed decreases in soil moisture. Projections for the twenty-first century show significant reductions in summer soil moisture content in the Mediterranean region (Lopez-Moreno and Beniston 2009). In particular, in the Mediterranean area, soil moisture content is expected to decline and saturation conditions are expected to be increasingly rare and restricted to periods in winter and spring (García-Ruiz et al. 2011). Changes in temperature and precipitation patterns and intensity will affect evapotranspiration and infiltration rates, and thus soil moisture. When depleted due to the lack of precipitation, increased evapotranspiration or increased runoff, soil moisture starts to constrain plant transpiration, crop growth and thus, livestock and food production (Lopez-Moreno and Beniston 2009).

We collected data on soil moisture content in the five agro-climatic zones in Egypt in 2015. These data were obtained from National Centers for Environmental Prediction; Climate Forecast System Reanalysis (NCEP/CFSR). Regular monthly means products with $0.5 \times 0.5°$ resolution for the soil layer between 0 and 10 cm were collected. These values were compared with the value of RCP6.0 produced by CCSM4 climate model in 2030.

Figure 7.3 showed that, in 2015, soil moisture content in the layer of 0–10 cm were the highest in winter months and the lowest in summer months. The figure also showed that, in 2030, soil moisture content will be almost the same during the winter and the summer months. This result implied that rising soil temperature in

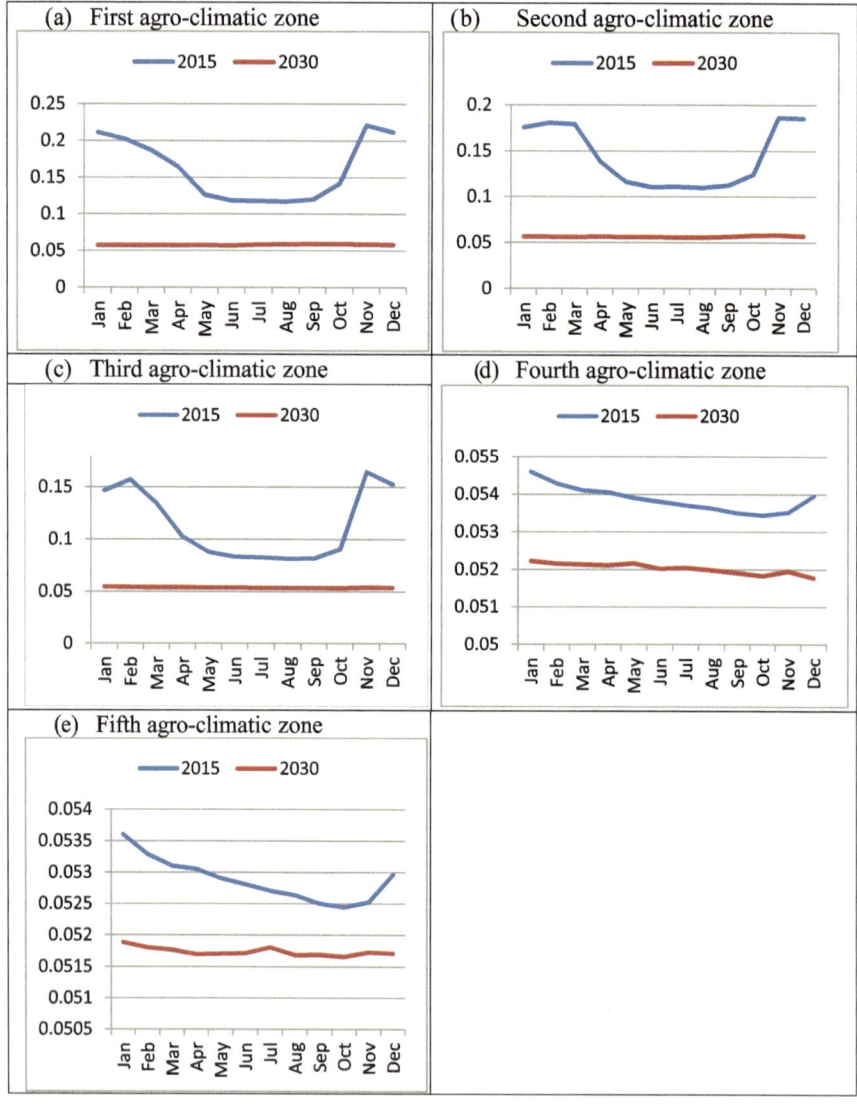

Fig. 7.3 Comparison between soil moisture content in 2015 and 2030 in the agro-climatic zones of Egypt

2030, could lead to increase soil evaporation and can reduce soil moisture content to a very low level. These results were true for the five agro-climatic zones in Egypt.

Effect of Climate Change on Evapotranspiration

Climate change is expected to increase reference evapotranspiration (ETo) due to higher temperature, higher solar radiation, and higher wind speed (Abtew and Melesse 2013), which will affect the hydrological system and water resources (Shahid 2011). Previous research in Egypt on the effect of climate change on ETo values revealed that temperature rise by 1 °C may increase ETo rate by about 4–5% (Eid 2001). Furthermore, Attaher et al. (2006), Khalil (2013) used SERS scenarios to calculate ETo values in the future and indicated that it will increase under climate change compared to current climate. Whereas, Ouda et al. (2011) developed prediction equations to calculate total water requirements needed to support irrigation in Egypt in 2025 and they found that an increase by 33% in water required for irrigation is expected to occur as a result of temperature increase by 2 °C and population growth. Ouda et al. (2016) indicated that evapotranspiration values will increase by an average of 9%, on a national level, in 2030, where it will be lower in north Egypt, compared to the middle and south of Egypt. Using their data, we calculated the increase in evapotranspiration in 2030 for each of the agro-climatic zones. An increase by 2, 3, 7, 12, and 14% in the first, second, third, fourth, and fifth agro-climatic zones, respectively, were found.

Effect of Climate Change on Crops Water Requirements

The BISm model (Snyder et al. 2004) was used to calculate ETo values (mm/day) in the studied five agro-climatic zones in 2030 using projected data by the RCP6.0 scenario resulted from CCSM4 model. The same sowing date set for crops cultivated in 2014/15 growing season was used in the BISm model, but harvest date was assumed to be earlier 3–5 days. Table 7.1 showed the expected planting and harvest dates in 2030.

Projected water consumptive use for the studied crops in each agro-climatic zone was calculated using BISm model and presented in Table 7.2.

Water requirements in 2030 for selected crops at each zone will be calculated under surface irrigation in the old lands of the Nile Delta and Valley using 60% application. In the new lands, water requirements for selected crops at each zone will be calculated under irrigation systems, namely sprinkler with 75% application efficiency and drip systems with 85% application efficiency.

Zohry and Ouda (2016a) indicated that climate change will increase water requirements for many crops in 2030, in comparison with its value in 2012. For

Table 7.1 Planting and harvest dates of selected crops in the prevailing cropping pattern

Crop	Planting date	Harvest date
Wheat	November 15	April 15
Faba bean	October 25	March 22
Clover	October 15	March 28
Onion	November 15	May 12
Tomato	October 1	February 27
Potato	November 1	March 29
Sugar beet	October 15	April 9
Cotton	March 15	August 26
Rice	May 15	September 11
Maize	May 15	August 26
Soybean	May 15	August 20
Sunflower	May 15	August 10
Tomato	May 1	August 26
Citrus	February 15	February 14
Olive	February 15	February 14
Grape	February 15	December 1

Table 7.2 Projected water consumptive use (mm) for the studied crops in the five agro-climatic zones in 2030

Crop	Zone 1	Zone 2	Zone 3	Zone 4	Zone 5
Wheat	384	390	433	473	501
Faba bean	348	361	376	421	451
Clover	572	600	626	695	742
Onion	683	746	779	855	901
Tomato	350	353	378	420	452
Potato	208	214	220	241	260
Sugar beet	547	585	609	665	–
Cotton	860	879	1005	1218	–
Rice	748	760	886	–	–
Maize	584	593	696	822	881
Soybean	576	604	608	–	–
Sunflower	511	520	611	721	773
Tomato	674	683	798	943	1008
Sugarcane	–	–	–	2200	2498

example, in the second agro-climatic zone, water requirements for wheat will increase by 4%. Water requirements for clover, sugar beet, faba bean, and winter tomato will increase by 7, 5, 3, and 8%, respectively. Furthermore, water requirements for the summer crops in the same zone will increase for maize, rice, soybean, sunflower and cotton, and summer tomato by 9, 9, 6, 6, 7, and 11%, respectively.

In the fourth agro-climatic zone, the increase in water requirements is expected to rise, where for wheat it will increase by 12%, whereas it will increase by 14% for clover, and by 17% for faba bean. Water requirements for maize will increase by 12% and by 11% for summer tomato. Regarding for sugarcane, its water requirements will increase by 13% (Zohry and Ouda 2016b).

Conclusion

Projected values of soil temperature on the level of each agro-climatic zone in 2030 revealed that there will be an increasing trend in its value in 2030, compared to its counterpart values in 2015. The highest differences will occur in the summer season from May to September. Whereas, in the winter season, there will be small differences between soil temperature in 2015 and 2030 from September to December, then the difference becomes higher from January to April. This, in turn, can result in increased water requirements for cultivated crops.

Furthermore, soil moisture content in the layer of 0–10 cm in 2030 will be reduced to the degree that it will have the same value in the winter as the summer due to increase in soil evaporation, which reduces soil moisture content to a very low level. These results were true for the five agro-climatic zones in Egypt.

Thus, the above consequences, in addition to the increase in the climatic elements will result in the expected increase in crops water requirements.

References

Abtew W, Melesse A (2013) Climate change and evapotranspiration. In Evaporation and evapotranspiration: measurements and estimations, Springer Science Business Media Dordrecht. doi:10.1007/978-94-007-4737-113

Attaher S, Medany M, AbdelAziz AA, El-Gendi A (2006) Irrigation-water demands under current and future climate conditions in Egypt. The 14th Annual Conference of the Misr Society of Agricultural. Engineering, pp 1051–1063

Calanca P, Roesch A, Jasper K, Wild M (2006) Global warming and the summertime evapotranspiration regime of the Alpine region. Clim Change 79:65–78. doi:10.1007/s10584-006-9103-9

Döll P (2002) Impact of climate change and variability on irrigation requirements: a global perspective. Clim Change 54:269–293

Eid H (2001) Climate change studies on Egyptian Agriculture. Soils, Water and Environment research institute SWERI ARC, Ministry of Agriculture, Giza, Egypt

Fujino J, Nair R, Kainuma M, Masui T, Matsuoka Y (2006) Multi-gas mitigation analysis on stabilization scenarios using AIM global model. Multi-gas mitigation and climate policy. Energy J Spec 27:343–353

García-Ruiz JM, López-Moreno JI, Vicente-Serrano SM, Lasanta–Martínez T, Beguería S (2011) Mediterranean water resources in a global change scenario. Earth-Sci Rev 105(3–4):121–139. doi:10.1016/j.earscirev.2011.01.006

Hijioka Y, Matsuoka Y, Nishimoto H, Masui M, Kainuma M (2008) Global GHG emissions scenarios under GHG concentration stabilization targets. J Glob Environ Eng 13:97–108

Holden N, Brereton A (2003) Potential impacts of climate change on maize production and the introduction of soybean in Ireland. Irish J Agric Food Res 42:1–15

IPCC (2013) Summary for policymakers. In: Climate change. The Physical Science Basis. Contribution of Working Group I to the Fifth Assessment Report of the Intergovernmental Panel on Climate Change. In: Stocker TF, Qin D, Plattner GK, Tignor M, Allen SK, Boschung J, Nauels A, Xia Y, Bex V, Midgley PM (eds) Cambridge University Press, Cambridge, United Kingdom and New York, NY, USA

Karmakar R, Das I, Dutta D, Rakshit A (2016) Potential effects of climate change on soil properties: a review. Sci Int 4(2):51–73

Khalil AA (2013) Effect of climate change on evapotranspiration in Egypt. Researcher 51:7–12

Kimball BA, Kobayashi K, Bindi M (2002) Responses of agricultural crops to free-air CO_2 enrichment. Adv Agron 77:293–368

Lopez-Moreno JI, Beniston M (2009) Daily precipitation intensity projected for the 21st century: seasonal changes over the Pyrenees. Theoret Appl Climatol 95:375–384. doi:10.1007/s00704-008-0015-7

Morsy M (2015) Use of regional climate and crop simulation models to predict wheat and maize productivity and their adaptation under climate change. Ph.D. thesis. Faculty of Science Al-Azhar University

Onol B, Semazzi FHM, Unal YS, Dalfes HN (2006) Regional climatic impacts of global warming over the eastern mediterranean. International Conference on Climate Change and the Middle East Past, Present and Future, pp 20–23 November 2006, Istanbul Technical University, Turkey

Ouda SA, Khalil F, El Afendi G, Abd El-Hafez SA (2011) Prediction of total water requirements for agriculture in the Arab world under climate change. Proceeding of the 15th International Conference on Water Technology, Egypt

Ouda S, Noreldin T (2017) Evapotranspiration data to determine agro-climatic zones in Egypt. J Water Land Dev 32(I–III):79–86

Ouda S, Noreldin T, Hosny M (2016) Evapotranspiration under changing climate. In: Major crops and water scarcity in Egypt. Springer Publishing House, pp 1–22

Ren J, Shen Z, Yang J, Zhao J, Yin J (2014) Effects of temperature and density on hydraulic conductivity of silty clay under infiltration of low-temperature water. Arab J Sci Eng 39:461–466. doi:10.1007/s13369-013-0849-x

Rustad LE, Huntington TG, Boone RD (2000) Controls on soil respiration: implications for climate change. Biogeochemistry 48:1–6

Seneviratne SI, Corti T, Davin EL, Hirschi M, Jaeger EB, Lehner I, Orlowsky B, Teuling AJ (2010) Investigating soil moisture–climate interactions in a changing climate: a review. Earth Sci Rev 99(3–4):125–161

Shahid S (2011) Impact of climate change on irrigation water demand of dry season boro rice in Northwest Bangladesh. Clim Change 105:433–453. http://dx.doi.org/10.1007/s10584-010-9895-5

Snyder RL, Orang M, Bali K, Eching S (2004) Basic Irrigation Scheduling (BIS). http://www.waterplan.water.ca.gov/landwateruse/wateruse/Ag/CUP/Californi/Climate_Data_010804.xls

Solomon S, Qin D, Manning M, Chen Z, Marquis M, Averyt K, Tignor M, Miller H (eds) (2007) Climate change 2007: the physical science basis. Contribution of working group I to the fourth assessment report of the intergovernmental panel on climate change. Cambridge University Press, Cambridge, United Kingdom and New York, NY, USA, p 996

Zohry AA, Ouda S (2016a) Crops intensification to face climate induced water scarcity in Nile Delta region. In: Management of climate induced drought and water scarcity in Egypt: Unconventional Solutions. Springer Publishing House

Zohry AA, Ouda S (2016b) Upper Egypt: Management of high water consumption crops by intensification. In: Management of climate induced drought and water scarcity in Egypt: Unconventional Solutions. Springer Publishing House

Chapter 8
Cropping Pattern to Face Climate Change Stress

Samiha A.H. Ouda and Abd El-Hafeez Zohry

Abstract The objective of this chapter is to quantify the effect of climate change in 2030 on the prevailing cropping pattern, in terms of increasing water requirements of the crops, which could result in the reduction of cultivated area. Furthermore, testing the effect management practices on the cultivated area of the copping pattern was also done in the five agro-climatic zones of Egypt. Our results indicated that increasing water requirements for the cultivated crops in each agro-climatic zone in 2030 will result in decrease in the cultivated area of these crops. The analysis showed that the cultivated area in the prevailing cropping pattern will be reduced in 2030 by 13%, compared to its counterpart value in 2014/15. A cropping pattern was suggested, where changing cultivation methods to raised beds will increase the cultivated area by 9%, compared to its counterpart value in 2014/2015. Furthermore, another cropping pattern was suggested, where cultivation on raised beds and polycropping will increase the cultivated area by 32%. Thus, the suggested management will overcome the stress resulted from climate change on the prevailing cropping pattern in the future.

Keywords Heat stress · Raised beds cultivation · Polycropping · Intercropping techniques

Introduction

Climate change is one of the overwhelming environmental threats that are defined as a long-term alteration in the global weather patterns, including temperature, precipitation, soil moisture, sea level, and storm activity (IPCC 2007). Most recent assessments of the effect of climate change on arid and semiarid regions concluded that these areas are highly vulnerable to climate change. The projected climatic changes will be among the most important challenges for agriculture in the twenty-first century, especially for developing countries and arid regions (IPCC 2007). The Intergovernmental Panel on Climate Change (IPCC 2001) stated that "water and its availability and quality, will be the main pressures on, and issues for, societies and the environment under climate change". Accordingly, it is expected

© The Author(s) 2018
S.A.H Ouda et al., *Cropping Pattern Modification to Overcome
Abiotic Stresses*, SpringerBriefs in Water Science and Technology,
https://doi.org/10.1007/978-3-319-69880-9_8

that climate change will induce disruption in food production systems in both irrigated and rainfed areas. The risks associated with agriculture and climate change arise out strong complicated relationships between agriculture and the climate system, plus the high reliance of agriculture on finite natural resources (IPCC 2013).

The response of crops to the different weather variables is quite complex and difficult to describe. If one of the variables is limiting (for example, temperatures that are too hot or too cold), then the effects of solar radiation or precipitation do not greatly affect the crop. When none of the variables is limiting, the crop will respond to the variable that is farthest from the optimum for that variable (Hollinger and Angel 2014). The interannual, monthly, and daily distribution of weather variables, such as temperature, radiation, precipitation, water vapor pressure, and wind speed affects a number of physical, chemical, and biological processes that drive the productivity of agricultural (IPCC 2007). Such climate change is a potential consequence of releasing the greenhouse gases (GHGs) that accumulated in the atmosphere, resulting in global warming (El Massah and Omran 2014). Furthermore, climate change is expected to increase potential evapotranspiration due to higher temperature, solar radiation, and wind speed (Abtew and Melesse 2013), which will affect the hydrological system and water resources (Shahid 2011).

Thus, the objective of this chapter is to quantify the effect of climate change in 2030 on the prevailing cropping pattern, in terms of increasing water requirements of the crops, which will result in the reduction of cultivated area. Furthermore, testing the effect management practices on the cultivated area of the copping pattern was also done in the five agro-climatic zones of Egypt. Figure 8.1 shows the five agro-climatic zones developed by Ouda and Noreldin (2017).

Heat Stress Resulted from Climate Change

Air temperature is the main variable that determines when a crop will grow. It also determines, along with precipitation and solar radiation, how well a crop will grow and how fast it will develop. Heat stress mostly occurs in the summer when temperatures approach or exceed 32 °C. Heat stress affects plants because as temperature increases, respiratory reaction rates speed up, using more of the photosynthetic compounds manufactured in a day (Gardner et al. 1985). Moreover, with elevated maximum temperature, especially temperatures that exceed 38 °C, plants require more water to maintain optimum water content in their tissues. If the soil cannot meet the additional water requirement, heat stress is compounded by an added water stress (Hollinger and Angel 2014). Increased temperatures caused by climate change may accelerate the rate at which plants release CO_2 in the process of respiration, resulting in less than optimal conditions for net growth (Gardner et al. 1985). When temperatures exceed the optimal for biological processes, crops often respond negatively with a steep drop in net growth and yield. Another important

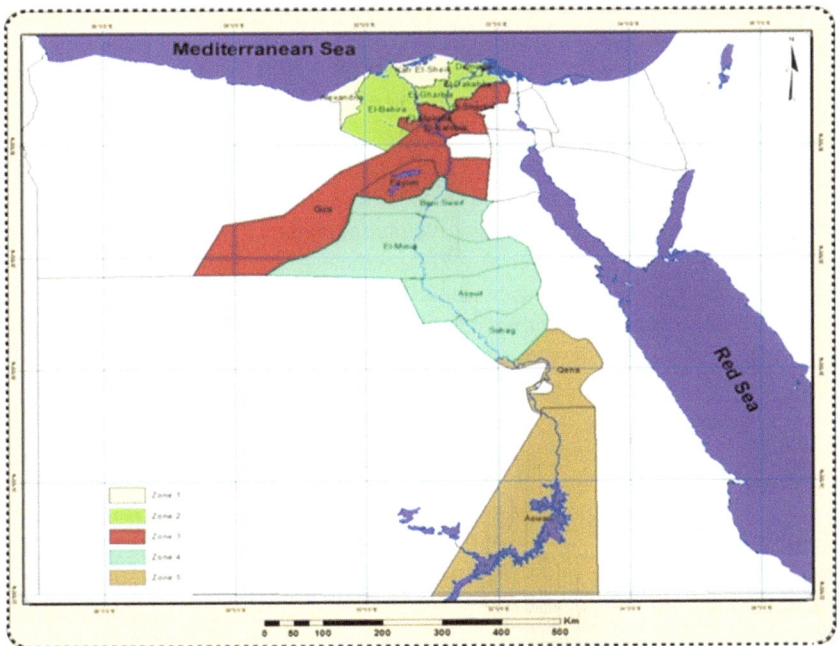

Fig. 8.1 Map of agro-climatic zones of Egypt using 10 years of ETo values

effect of high temperature is accelerated physiological development, resulting in hastened maturation and reduced yield (Vu et al. 2001).

Wheat plants are affected by heat stress to varying degrees at different phenological stages (Slafer and Satorre 1999), but heat stress during the reproductive phase is more harmful than during the vegetative phase due to the direct effect on grain number and its dry weight (Wollenweber et al. 2003). Pre-anthesis and post-anthesis high temperature may have huge impacts upon wheat growth through reduction in photosynthetic efficiency (Yang et al. 2011).

Heat stress negatively affects floral stages in faba bean, where flowers are the most affected during initial green-bud stages (Bishop et al. 2016). Yield and pollen germination of flowers present before heat stress showed threshold relationships to stress, with lethal temperatures between 28 and 32 °C, whereas whole plant yield showed a linear negative relationship to stress with high plasticity in yield allocation (Bishop et al. 2016).

Maize plants become susceptible to high temperatures after reaching eight-leaf stage to seventeen-leaf stage (Chen et al. 2010), which has significant impact on plant growth, architecture, ear size, and kernel numbers (Farré and Faci 2006).

Reddy et al. (2002) concluded that cotton yield was decreased by 9% under climate change conditions. The rate of plant growth and development was higher in the future because of enhanced metabolic rates at higher temperatures combined with increased carbon availability. They also stated that since most of the days with average temperatures above 32 °C will likely occur during the reproductive phase, irrigation will be needed to satisfy the high water demand, and this reduces boll abscission by lowering canopy temperatures. Furthermore, Gwimbi (2009) reported that cotton production levels will be declined as precipitation decreased and temperatures increased under climate change conditions.

Rice is relatively tolerant to high temperatures during the vegetative phase but highly susceptible during the reproductive phase, particularly at the flowering stage (Jagadish et al. 2010). Matsui et al. (2001) indicated that 50% of spikelet sterility was recorded by 3 °C difference in critical temperature between the tolerant genotype. Temperature higher than 35 °C for more than 3 days during reproductive stages can affect pollen development and pollination, resulting in decreased seed setting and production. Heat stresses occurred during the reproductive organ developing and flowering stages resulted in severe loss of production (Zou et al. 2009). Booting and flowering stages are most sensitive to high temperature, which may sometimes lead to complete sterility.

Sugarcane exposed to high temperatures is likely to negatively affect sprouting and emergence of sugarcane (Rasheed et al. 2011). Poor emergence of sugarcane will result in significantly low plant population. In addition, temperatures above 32 °C result in short internodes, increased number of nodes, and lower sucrose (Bonnett et al. 2006). Furthermore, Chandiposha (2013) revealed that high temperatures especially at night usually result in more flowering of sugarcane. Flowering in sugarcane ceases growth of leaves and internodes, which reduces sugarcane and sucrose yields. At tillering stage, the crop favored higher minimum temperature (about 26.2 °C), whereas temperatures above 38 °C make sugarcane growth seize (Bonnett et al. 2006).

Effect of Climate Change on Crops Productivity

A range of valuable national studies has been carried out and published in Egypt concerning the potential vulnerability of several crops under expected climate change in the future, such as wheat, maize, and cotton.

Regarding wheat in the first agro-climatic zone, Hassanein et al. (2012) stated that under climate change conditions, wheat productivity will be reduced by 12–27% averaged over three cultivars, compared to 2009/2010 growing season. In the same zone, Kassem (2016) indicated that climate change will reduce wheat productivity by 23–27%, compared to its productivity in 2014/2015 growing season.

In the second agro-climatic zone, the productivity of wheat planted in clay soil under sprinkler system is expected to be reduced by 21% as an average over four cultivars, compared to wheat yield grown in 2009/2010 growing season (Abdrabbo et al. 2013). Furthermore, farmer's application of irrigation water in sandy soil in the same zone using sprinkler system is characterized by large applied amount. This practice will increase wheat yield losses under climate change. Wheat productivity is expected to be reduced by 30% for farmer practice (Ouda et al. 2010) and by 38% under fertigation practice (Ibrahim et al. 2012). In salt-affected soil in the same zone and under surface irrigation, wheat yield was reduced by 40% for farmer's practice (Noreldin et al. 2013).

Moreover, in the third agro-climatic zone, Khalil et al. (2009) indicated that the productivity of wheat in clay soil is expected to decrease by 40% as an average of three cultivars. At the same zone, wheat was grown in silty clay soil under sprinkler system, and its productivity will be reduced by 21% as an average over four cultivars (Noreldin et al. 2012). Morsy (2015) indicated that wheat productivity will be reduced by 5–15% in the first, second, and third agro-climatic zones, respectively.

In the fourth agro-climatic zone and in 2009/2010 growing season, Hassanein et al. (2012) found that wheat productivity will be decreased by 11–31% as an average over three cultivars.

With respect to maize grown in the second agro-climatic zone, its productivity in clay soil under drip irrigation expected to be reduced by 25% (Ouda et al. 2012). In the same zone and in sandy soil under drip irrigation, maize yield will be reduced by 41% for farmer's irrigation and by 36% when irrigation was applied using 1.0 ETc and fertigation application in 80% of irrigation time (Ouda et al. 2014).

In the third agro-climatic zone, maize productivity will be reduced by 55% in clay soil under surface irrigation, with deterioration in water productivity (Ouda et al. 2009). In the same zone, maize yield was grown in silty clay soil under drip irrigation; its yield was reduced by 27% as an average of four hybrids (Ouda et al. 2012). Morsy (2015) indicated that maize productivity will be reduced in the first, second, and third agro-climatic zones by the percentage between 7 and 19%.

Finally, for cotton grown in the second agro-climatic zone, reduction in its yield under climate change scenarios will be 22–29% for farmer irrigation. Application of required irrigation amount under climate change scenarios will reduce cotton yield losses to 20–25%, whereas raised beds cultivation reduced cotton yield losses by 17–22%, compared to farmer irrigation amount (Ouda et al. 2013).

Effect of Climate Change on the Soil

Soils are important for food security and climate change has the potential to threaten food security through its effects on soil properties and processes (Pimentel 2006). Climate is one of the most important factors affecting the formation of soil with important implications for their development, use, and management perspective (Karmakar et al. 2016). Increased temperature is likely to have a negative effect on C allocation to the soil, leading to reductions in soil organic C and creating a positive feedback in the global C cycle, where increased temperatures lead to increased CO_2 release from soils to the atmosphere and consequently lead to more increases in temperature (Wan et al. 2011). A decline in soil organic matter levels leads to a decrease in soil aggregate stability and infiltration rates, and increase susceptibility to compaction, run-off, and erosion (Gorissen et al. 2004).

Furthermore, increased salinization and alkalization would occur in areas where evaporation increased or rainfall decreased (Varallyay 1994). Transient salinity increases as capillary rise dominates, bringing salts into the root zone on sodic soils. Increased subsoil drying increases the concentration of salts in the soil solution. Conversely, the severity of saline scalds due to secondary salinization may decline as groundwater levels fall in line with reduced rainfall. This development could have significant impacts on large areas in the semiarid zones (Karmakar et al. 2016). Moreover, salinization can also be a consequence of expected climate change, as the rise of sea level and seawater intrusion (Varallyay 1994).

Climate change can affect soils build up due to an increase of the evapotranspiration through the increase in air temperature (Várallyay 1994). The degradation caused by climate change was summarized by Várallyay (2010) in increased soil erosion; therefore, it should be balanced by the increasing soil conservation effect of more dense and permanent vegetation. Higher rate of evapotranspiration will increase capillary transport of water and solutes from the groundwater to the root zone.

Under climate change, crop rotation can play an important role in the reduction of GHGs fluxes from the soil by more efficient management of carbon and nitrogen flows in agricultural ecosystems (Cerrie et al. 2004).

Effect of Climate Change on Soil Temperature and Seed Germination

Under optimum soil and irrigation conditions, soil temperature is one of the major environmental factors that influence not only the proportion of seeds that germinate but also the rate of emergence and subsequent establishment (Prasad et al. 2006). Temperature plays a major role in determining the periodicity of seed germination and the distribution of species (Guan 2009). Seed germination is a complex process involving many individual reactions and phases, each of which is affected by temperature (Canossa 2008). The effect on germination can be expressed in terms of cardinal temperature: that is minimum, optimum, and maximum temperatures at which germination will occur (Prasad et al. 2006), which can determine some of the ecological limitations for the geographic distribution of the species (Flores and Briones 2001). The minimum temperature is sometimes difficult to define since germination may actually be proceeding but at such a slow rate that determination of germination is often made before actual germination is completed. The optimum temperature may be defined as the temperature giving the greatest percentage of germination in the shortest time. The optimum temperatures produce both the most rapid seed germination and plant growth (Hakansson et al. 2002). The maximum temperature is governed by the temperature at which denaturation of proteins essential for germination occurs. The optimum temperature for most seeds is between 15 and 30 °C. The maximum temperature for most species is between 30 and 40 °C (Prasad et al. 2006). Germination speed usually increases until the temperature reaches 30–35 °C (Roberts and Ellis 1989). Germination rate usually increases linearly with temperature, at least within a well-defined range, and declines sharply at higher temperatures (Alvarado and Bradford 2002).

Projected Cropping Pattern Under Climate Change

As it was stated before, the cultivated areas are expected to decrease in 2030 as a result of increase in water requirements for the cultivated crops. We calculated the expected reduction in the cultivated area of the cropping pattern of 2014/2015 in each agro-climatic zone. Table 8.1 shows the expected value of the total cultivated area in the old and new lands in each zone. The results indicated that the cultivated

Table 8.1 Expected cultivated area of the olds and new lands in 2030 in the agro-climatic zones of Egypt

	Zone 1	Zone 2	Zone 3	Zone 4	Zone 5
Winter crops					
Wheat	121,856	374,037	361,900	319,029	83,183
Faba bean	7141	17,552	4798	1957	905
Clover	53,240	219,128	183,448	100,908	11,790
Onion	15,855	49,603	33,995	27,589	1868
Tomato	2256	16,534	27,163	12,410	6197
Potato	4731	53,772	28,336	9094	216
Sugar beet	62,647	81,871	43,784	24,395	0
Others	22,233	178,993	79,268	31,756	130,793
Total	289,957	991,492	762,692	527,138	234,952
Summer crops					
Cotton	27,302	38,131	19,232	0	0
Rice	91,390	268,197	85,911	0	0
Maize	29,450	173,124	253,949	275,916	27,608
Soybean	357	703	104	10,564	0
Sunflower	98	3244	1231	892	0
Potato	2881	24,668	16,388	0	53
Tomato	16,252	31,388	15,772	7734	713
Sugarcane	0	0	0	17,844	93,921
Fruit trees	6333	225,299	162,228	40,953	20,752
Others	109,380	227,474	217,227	73,404	28,602
Total	283,443	992,230	772,042	427,305	171,648
Grand total	573,400	1,983,722	1,534,734	954,443	406,600

Table 8.2 Expected cultivated area of total old and new lands in 2030 in Egypt

	Total old lands (ha)	Total new lands (ha)	Total cultivated area (ha)
Winter crops			
Wheat	1,065,565	194,440	1,260,005
Faba bean	18,448	13,905	32,353
Clover	523,406	45,108	568,514
Onion	106,571	22,340	128,910
Tomato	26,452	38,108	64,559
Potato	78,929	17,219	96,148
Sugar beet	162,393	50,304	212,697
Others	106,985	336,059	443,043
Total	2,088,747	717,483	2,806,230
Summer crops			
Cotton	77,520	9784	84,665
Rice	413,328	32,171	445,499
Maize	691,691	68,356	760,046
Soybean	11,380	348	11,728
Sunflower	2634	2831	5465
Potato	36,724	7888	43,989
Tomato	30,700	41,159	71,860
Sugarcane	98,976	12,789	111,765
Fruit trees	184,107	271,458	455,565
Others	571,684	84,403	656,087
Total	2,118,744	531,187	2,646,669
Grand total	4,207,491	1,248,670	5,452,899

area is expected to be decreased by 19, 8, 14, 15, and 23% for the first, second, third, fourth, and fifth agro-climatic zones, respectively, in comparison with the cultivated area in the cropping pattern of 2014/2015 (Chap. 3: Tables 3.3, 3.5, 3.7, 3.9, and 3.11).

Table 8.2 presented the expected total national cultivated area in the old and new lands in 2030. The table showed that the total cropped area will be reduced in 2030 by 13%, in comparison with the cropped area in the cropping pattern of 2014/2015 (Chap. 3: Table 3.12).

Increasing Resilience of the Cropping Pattern to Climate Change

1. Cultivation on raised beds

In an attempt to reduce the vulnerability of the national cropping pattern to climate change, we assessed the effect of changing cultivation method to raised beds (can

Table 8.3 Potential added cultivated area (ha) in the old lands as a result of cultivation on raised beds in the agro-climatic zones of Egypt in 2030

	Zone 1	Zone 2	Zone 3	Zone 4	Zone 5
Winter crops					
Wheat	25,144	75,292	83,047	71,831	11,083
Faba bean	1517	2131	1030	489	60
Clover	11,803	49,236	43,892	23,823	2099
Onion	3354	8905	5690	4234	122
Tomato	558	854	4579	1221	283
Potato	1006	10,910	7889	2557	1
Sugar beet	16,974	14,476	8920	5643	0
Others	7404	9475	70,600	4508	518
Total	67,759	171,279	225,649	114,304	14,165
Summer crops					
Cotton	7644	9198	4374	0	0
Rice	25,467	69,814	21,829	0	0
Maize	7036	39,029	69,007	75,155	5752
Soybean	0	106	0	2988	0
Sunflower	7	306	202	231	0
Potato	52	5618	4544	0	0
Tomato	2158	2463	2593	1345	139
Sugarcane	0	0	0	4020	19,075
Fruit trees	1378	13,337	25,061	8314	4074
Others	5095	21,349	26,482	18,460	16,942
Total	48,836	161,222	154,092	110,512	45,981
Grand total	116,595	332,501	379,740	224,816	60,147

save 20% of the applied water to surface irrigation) on the national cropped area in 2030. Table 8.3 indicated that the added area as a result of irrigation water availability in 2030 will be lower, in comparison with its counterpart value in the cropping pattern of 2014/2015. The highest added area will exist in the third agro-climatic zone and the lowest added area will exist in the fifth agro-climatic zone. The added cultivated area was reduced by 6, 12, 12, 13, and 15% for the first, second, third, fourth, and fifth agro-climatic zones, respectively, in comparison with the added area in the cropping pattern of 2014/2015 (Chap. 5: Table 5.2).

2. **Cultivation on raised beds and implementing polycropping**

Table 8.4 indicated that cultivation on raised beds and implementing polycropping (intercropping systems and cultivation of middle crop between winter and summer crop, see the details in Chap. 5) can increase the potential added area. Thus, the added area under this practice will be lower than its counterpart values in 2015

Table 8.4 Potential added cultivated area (ha) in the old lands as a result of cultivation on raised beds and polycropping in the agro-climatic zones in Egypt in 2030

	Zone 1	Zone 2	Zone 3	Zone 4	Zone 5
Winter crops					
Wheat	65,839	171,519	154,438	94,686	26,913
Faba bean	19,320	66,913	51,514	22,200	21,506
Clover	11,803	49,236	43,892	23,823	2099
Short season clover	91,000	91,000	91,000	91,000	91,000
Onion	3354	8905	5690	4234	122
Tomato	0	0	0	0	0
Potato	1006	10,910	7889	2557	1
Sugar beet	0	0	0	0	0
Others	7404	9475	70,600	4508	518
Total	108,725	316,958	334,024	152,008	51,158
Summer crops					
Cotton	0	0	0	0	0
Rice	25,467	69,814	21,829	0	0
Maize	7492	15,434	24,824	32,582	5447
Soybean	0	0	0	0	0
Sunflower	8599	256,828	8297	9026	15,402
Sunflower (early)	6917	6917	6917	6917	6917
Potato	52	5618	4544	0	0
Tomato	0	0	0	0	0
Cowpea	20,683	123,458	178,626	193,766	19,325
Sugarcane	0	0	0	0	0
Fruit trees	0	0	0	0	0
Others	5095	21,349	154,092	18,460	16,942
Total	74,303	499,418	399,128	260,750	64,033
Grand total	183,029	816,376	733,152	412,758	115,192

(Chap. 5: Table 5.4) by 12, 15, 14, 15, and 21% for the first, second, third, fourth, and fifth agro-climatic zones, respectively. It is worth noting that the main crops in the suggested cropping systems, such as winter and summer tomato, sugar beet, cotton, soybean, sugarcane, and fruit trees, will not contribute to saved irrigation water because its saved irrigation water will be used to irrigate the middle crop, namely short season clover and early sunflower.

Table 8.5 presented the final cropping pattern in case of cultivation on raised beds only in 2030, where the total potential cropped area will reach 6,788,291 ha. Furthermore, in case of implementing polycropping in addition to cultivation on raised beds, the potential cropped area will reach 8,220,670 ha.

Table 8.5 Potential added cultivated area and potential total area as a result of cultivation on raised beds and as a result of cultivation on raised beds and polycropping in 2030

	Raised beds cultivation		Raised beds and polycropping	
	Added area (ha)	Total area (ha)	Added area (ha)	Total area (ha)
Winter crops				
Wheat	266,396	1,526,401	513,395	1,773,400
Faba bean	8089	37,580	181,454	213,807
Clover	243,550	699,366	130,853	726,417
Short season clover	–	–	455,000	455,000
Onion	51,762	151,215	22,305	153,041
Tomato	26,102	72,054	0	51,895
Potato	31,672	118,510	22,362	114,156
Sugar beet	70,408	258,710	0	196,394
Others	254,604	704,228	92,505	567,874
Total	952,584	3,568,065	1,417,873	4,251,984
Summer crops				
Cotton	23,855	108,520	0	90,836
Rice	117,109	562,608	117,109	610,628
Maize	499,503	956,025	85,778	849,480
Soybean	13,658	14,822	0	11,505
Sunflower	1639	6211	298,151	301,759
Sunflower (early)	–	–	34,583	34,583
Potato	10,889	54,826	10,214	55,606
Tomato	17,145	80,558	0	51,628
Cowpea	0	0	661,439	661,439
Sugarcane	134,859	134,859	0	111,765
Fruit trees	113,868	507,729	0	290,677
Others	239,985	794,067	215,938	898,780
Total	1,172,510	3,220,226	1,423,213	3,968,686
Grand total	2,125,094	6,788,291	2,841,086	8,220,670

Conclusion

Increasing water requirements for the cultivated crops in each agro-climatic zone in 2030 will result in decrease in the cultivated area of these crops. The analysis showed that the cultivated area in the prevailing cropping pattern will be reduced in 2030 by 13%, compared to its counterpart value in 2014/2015. A cropping pattern was suggested, where changing cultivation methods to raised beds will increase the cultivated area by 9%, compared to its counterpart value in 2014/2015. Furthermore, another cropping pattern was suggested, where cultivation on raised beds and polycropping will increase the cultivated area by 32%. Thus, the suggested management will overcome the stress resulted from climate change on the prevailing cropping pattern in the future.

References

Abdrabbo M, Ouda S, Noreldin T (2013) Modeling the effect of irrigation scheduling on wheat under climate change conditions. Nat Sci J 115:10–18

Abtew W, Melesse A (2013) Climate change and evapotranspiration. In: Evaporation and evapotranspiration: measurements and estimations. Springer Science Business Media, Dordrecht. doi:10.1007/978-94-007-4737-113

Alvarado V, Bradford KJ (2002) Hydrothermal time model of seed germination explains the cardinal temperatures for seed germination. Plant, Cell Environ 25:1061–1069

Bishop J, Potts SG, Jones HE (2016) Susceptibility of faba bean (*Vicia faba* L.) to heat stress during floral development and anthesis. J Agro Crop Sci 202:508–517

Bonnett GT, Hewitt ML, Glassop D (2006) Effects of high temperature on the growth and composition of sugarcane internodes. Aust J Agric Res 5710:1087–1095

Canossa RS (2008) Effect of temperature and light on joy weed (*Alternanthera tenella*) seed germination. Planta Daninha 26(4):745–750

Cerrie CC, Bernoux M, Cerrie CEP, Feller C (2004) Carbon cycling and sequestration opportunities in South America: the case of Brazil. Soil Use Manag 20:248–254

Chandiposha M (2013) Review: potential impact of climate change in sugarcane and mitigation strategies in Zimbabwe. Afr J Agric Res 8:2814–2818; Clowes MJ, Breakwell WL (1998) Zimbabwe sugarcane production manual. Zimbabwe Sugar Association, Chiredzi

Chen J, Xu W, Burke JJ, Xin Z (2010) Role of phosphatidic acid in high temperature tolerance in maize. Crop Sci 50:2506–2515

El Massah S, Omran G (2014) Would climate change affect the imports of cereals? The case of Egypt. Handbook of climate change adaptation. Springer, Berlin, pp 657–685

Farré I, Faci JM (2006) Comparative response of maize *Zea mays* L. and sorghum *Sorghum bicolor* L. Moench to deficit irrigation in a Mediterranean environment. Agric Water Manag 83:135–143

Flores J, Briones O (2001) Plant life-form and germination in a Mexican inter-tropical desert: effects of soil water potential and temperature. J Arid Environ 47:485–497

Gardner FP, Pearce RB, Mitchell RL (1985) Physiology of crop plants. Iowa State University Press, Ames

Gorissen A, Tietema A, Joosten NN, Estiarte M, Peñuelas J, Sowerby A, Emmett BA, Beier C (2004) Climate change affects carbon allocation to the soil in shrub lands. Ecosystems 7:650–661

Guan B (2009) Germination responses of *Medicago ruthenica* seeds to salinity, alkalinity, and temperature. J Arid Environ 73(1):135–138

Gwimbi P (2009) Cotton farmers' vulnerability to climate change in Gokwe District (Zimbabwe): impact and influencing factors. J Disaster Risk Stud 2(1):153–157

Hakansson I, Myrbeck A, Ararso E (2002) A review of research on seedbed preparation for small grains in Sweden. Soil Tillage Res 64:23–40

Hassanein MK, Elsayed M, Khalil AA (2012) Impacts of sowing date, cultivar, irrigation regimes and location on bread wheat production in Egypt under climate change conditions. Nat Sci 10 (12):141–150

Hollinger SE, Angel JR (2014) Weather and crops. In: Illinois agronomy handbook. University of Illinois, USA, pp 1–12

Ibrahim MM, Ouda SA, Taha A, El Afandi G, Eid SM (2012) Water management for wheat grown in sandy soil under climate change conditions. J Soil Sci Plant Nutr 12(2):195–210

IPCC (2001) Climate change 2001: impacts, adaptation, and vulnerability. Report of working group II of the Intergovernmental Panel on Climate Change (6th Session, Geneva)

IPCC (2013) Summary for policymakers. In: Stocker TF, Qin D, Plattner GK, Tignor M, Allen SK, Boschung J, Nauels A, Xia Y, Bex V, Midgley PM (eds) Climate change. The physical science basis. Contribution of working group I to the fifth assessment report of the Intergovernmental Panel on Climate Change. Cambridge University Press, Cambridge, and New York

IPCC Intergovernmental Panel on Climate Change (2007) Intergovernmental Panel on Climate Change fourth assessment report: climate change 2007. Synthesis Report. World Meteorological Organization, Geneva

Jagadish SVK, Muthurajan R, Oane R, Wheeler TR, Heuer S, Bennett J, Craufurd PQ (2010) Physiological and proteomic approaches to dissect reproductive stage heat tolerance in rice *Oryza sativa* L. J Exp Bot 61:143–156

Karmakar R, Das I, Dutta D, Rakshit A (2016) Potential effects of climate change on soil properties: a review. Sci Int 4(2):51–73

Kassem A (2016) Impact of climate change on wheat under water stress in North Nile Delta. Ph. D., Mansoura University

Khalil FA, Farag H, El Afandi G, Ouda SA (2009) Vulnerability and adaptation of wheat to climate change in Middle Egypt. In: Proceedings of the 13th International Conference on Water Technology. Egypt, 12–15 March

Matsui T, Omasa K, Horie T (2001) The difference in sterility due to high temperatures during the flowering period among japonica rice varieties. Plant Prod Sci 4:90–93

Morsy M (2015) Use of regional climate and crop simulation models to predict wheat and maize productivity and their adaptation under climate change. Ph.D. thesis, Faculty of Science, Al-Azhar University

Noreldin T, Abdrabbo M, Ouda S (2012) Increasing water productivity for wheat grown under climate change conditions. In: 10th International Conference of Egyptian Soil Science Society (ESSS) and 4th International Conference of Water Requirements & Metrology Dept., 5–7 November, Aameria, Egypt

Noreldin T, Ouda S, Abou Elenein R (2013) Development of management practices to address wheat vulnerably to climate change in North Delta. In: Proceedings of the 11th International Conference on Development of Dry lands. Beijing, China

Ouda S, Noreldin T (2017) Evapotranspiration data to determine agro-climatic zones in Egypt. J Water Land Dev 32(I–III):79–86

Ouda SA, Khalil FA, Yousef H (2009) Using adaptation strategies to increase water use efficiency for maize under climate change conditions. In: Proceedings of 13th International Conference on Water Technology. Egypt

Ouda SA, Sayed M, El Afandi G, Khalil FA (2010) Developing an adaptation strategy to reduce climate change risks on wheat grown in sandy soil in Egypt. In: Proceedings of 10th International Conference on Development of Dry lands. Egypt

Ouda S, Abdrabbo M, Noreldin T (2012) Effect of changing sowing dates and irrigation scheduling on maize yield grown under climate change conditions. In: 10th International Conference of Egyptian Soil Science Society (ESSS) and 4th International Conference of Water Requirements & Metrology Dept., 5–7 November, Aameria, Egypt (Cl 07)

Ouda S, Noreldin T, Abou Elenein R, Abd El-Baky H (2013) Adaptation of cotton crop to climate change in salt affected soil. In: Proceedings of the 11th International Conference on Development of Dry lands. Beijing, China

Ouda S, Taha A, Ibrahim M (2014) Increasing water productivity for maize grown in sandy soil under climate change conditions. Arch Agron Soil Sci 61(3):299–311

Pimentel D (2006) Soil erosion: a food and environmental threat. Environ Dev Sustain 8:119–137

Prasad PVV, Boote KJ, Thomas JMG, Allen LH Jr, Gorbet DW (2006) Influence of soil temperature on seedling emergence and early growth of peanut cultivars in field conditions. J Agron Crop Sci 192(3):168–177

Rasheed R, Wahid A, Farooq M, Hussain I, Basra SMA (2011) Role of proline and glycine betaine pretreatments in improving heat tolerance of sprouting sugarcane *Saccharum* sp. buds. Plant Growth Regul 65:35–45

Reddy KR, Prashant RD, Mearns LM, Boone M, Hodges HF, Richardson GL, Kakani VG (2002) Simulating the impacts of climate change on cotton production in the Mississippi delta. Clim Res 22:271–281

Roberts EH, Ellis RH (1989) Water and seed survival. Ann Bot 63:39–52

Shahid S (2011) Impact of climate change on irrigation water demand of dry season Boro rice in northwest Bangladesh. Clim Change 105:33–453. http://dx.doi.org/10.1007/s10584-010-9895-5

Slafer GA, Satorre EH (1999) Wheat: ecology and physiology of yield determination. Haworth Press Technology and Industrial. ISBN 1560228741

Várallyay G (1994) Climate change, soil salinity and alkalinity. In: Rounsevell MDA, Loveland PJ (eds) Soil responses to climate change. Springer, Heidelberg, pp 39–54. ISBN 978-3-642-79220-5

Várallyay G (2010) The impact of climate change on soils and on their water management. Agron Res 8(Special Issue II):385–396

Vu JCV, Gesch RW, Pennanen AH, Allen LH Jr, Boote KJ, Bowes G (2001) Soybean photosynthesis, Rubisco, and carbohydrate enzymes function at supraoptimal temperatures in elevated CO_2. J Plant Physiol 158:295–307

Wan Y, Lin E, Xiong W, Li Y, Guo L (2011) Modeling the impact of climate change on soil organic carbon stock in upland soils in the 21st century in China. Agric Ecosyst Environ 141:23–31

Wollenweber B, Porter JR, Schellberg J (2003) Lack of interaction between extreme high-temperature events at vegetative and reproductive growth stages in wheat. J Agron Crop Sci 189:142–150

Yang F, Jørgensen AD, Li H, Søndergaard I, Finnie C, Svensson B, Jiang D, Wollenweber B, Jacobsen S (2011) Implications of high-temperature events and water deficits on protein profiles in wheat (*Triticum aestivum* L. cv. Vinjett) grain. Proteomics 11:1684–1695

Zou J, Liu AL, Chen XB, Zhou XY, Gao GF, Wang WF, Zhang XW (2009) Expression analysis of nine rice heat shock protein genes under abiotic stresses and ABA treatment. J Plant Physiol 166:851–861

Chapter 9
Cropping Index Under Proposed Cropping Patterns

Abd El-Hafeez Zohry and Samiha A.H. Ouda

Abstract The objective of this chapter is to compare between the cropping index under the prevailing cropping pattern and the suggested cropping patterns to overcome water scarcity, salinity stress, and climate change stress in the five agro-climatic zones in Egypt. One common measurement used in Egypt to measure how cropping pattern can satisfy the needs of the population for food is cropping index. It is calculated by dividing the cropped area by the actual cultivated area multiply by 100%. The cropped area is calculated by adding the winter cultivated area to the summer cultivated area and any other additional cultivated area either early winter or early summer. Our results showed that the highest cultivated area was found in the second agro-climatic zone and the lowest cultivated area was found in the fifth agro-climatic zone. The highest value of cropping index was found in the second agro-climatic zone under prevailing cropping pattern, under cropping pattern that increases food security, and under cropping pattern that faces water scarcity. Furthermore, in the cropping pattern that faces salinity and cropping pattern that faces climate change, the highest value of cropping index was found in the third agro-climatic zone. The highest percentage of increase in the cropping index value was found in the fourth agro-climatic zone.

Keywords Cropping index · Agro-climatic zones of Egypt · Cropped area

Introduction

Cropping pattern is the yearly sequence and spatial arrangement of crops or of crops and fallow on a given area. It indicates the proportion of area under different crops at a point of time (Madari and Shekadar 2015). It should provide enough food for the family, fodder for cattle, and generate sufficient cash income for domestic and cultivation expenses. Cropping pattern is evolved based on climate, water availability, and soil for efficient use of available natural resources.

One common measurement used in Egypt to measure how cropping pattern can satisfy the needs of the population for food is cropping index. It is calculated by dividing the cropped area by the actual cultivated area and multiplying by 100%.

© The Author(s) 2018
S.A.H Ouda et al., *Cropping Pattern Modification to Overcome Abiotic Stresses*, SpringerBriefs in Water Science and Technology,
https://doi.org/10.1007/978-3-319-69880-9_9

103

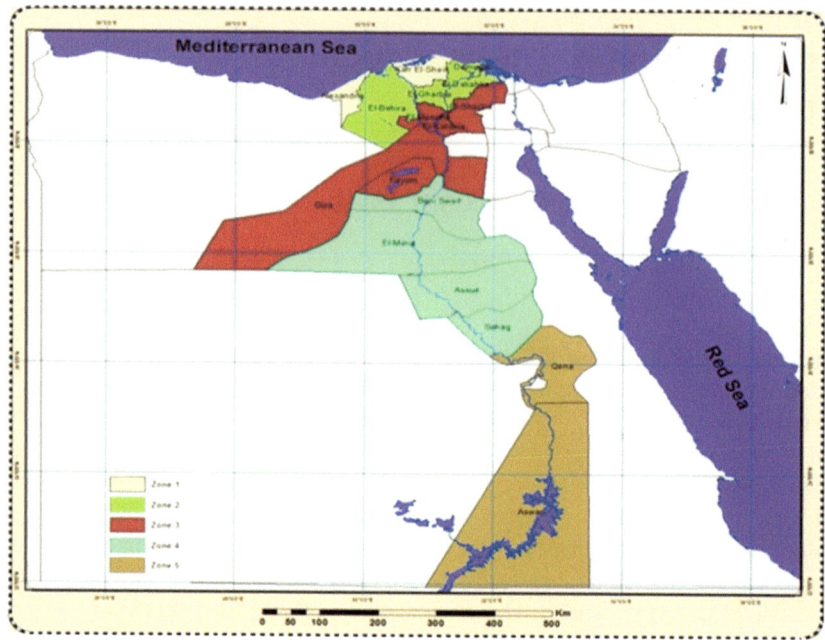

Fig. 9.1 Map of agro-climatic zones of Egypt using 10 years of ETo values

The cropped area is calculated by adding the winter cultivated area to the summer cultivated area and any other additional cultivated area either early winter or early summer.

The objective of this chapter is to compare between the cropping index under the prevailing cropping pattern and the suggested cropping patterns to overcome water scarcity, salinity stress, and climate change stress in the five agro-climatic zones in Egypt. Figure 9.1 shows the five agro-climatic zones developed by Ouda and Noreldin (2017).

Cropping Index for the Prevailing Cropping Pattern

The cropping pattern in Egypt depends on five main crops, namely wheat, maize, clover, cotton, and sugarcane. Wheat and maize for flour production, clover for feed, cotton for fiber and edible oil, and sugarcane for sugar production. Other crops exist in the Egyptian cropping pattern, such as cereals, legumes, fibers, forages, vegetables, and fruit crops. Table 9.1 presented the total cultivated area (cropped area) in each agro-climatic zone and its cropping index (Shafsak and El-Debaby 1975). The term "others" in Table 9.1 refers to the cultivated area of a large number of cereals, legumes, fibers, forages, vegetables, and fruit crops gathered together in

Table 9.1 Total cultivated area and cropping index values (CI, %) in the agro-climatic zones of Egypt under the prevailing cropping pattern

	Zone 1	Zone 2	Zone 3	Zone 4	Zone 5	Total
Winter crops						
Wheat	131,028	402,191	389,140	343,042	89,444	1,354,844
Faba bean	7597	18,672	5105	2082	963	34,418
Clover	58,505	240,800	201,591	110,888	12,956	624,741
Onion	18,225	57,015	39,075	31,711	2147	148,173
Tomato	2452	17,972	29,525	13,489	6736	70,173
Potato	5033	57,205	30,144	9674	230	102,285
Sugar beet	68,094	88,990	47,591	26,517	0	231,193
Others	87,344	198,881	213,357	35,285	145,410	680,278
Total	378,276	1,081,727	955,527	572,688	257,886	3,246,104
Summer crops						
Cotton	31,381	43,829	22,105	3033	0	100,349
Rice	103,853	304,770	97,627	0	0	506,249
Maize	36,358	213,734	313,517	340,637	34,084	938,329
Soybean	430	847	125	12,728	0	14,130
Sunflower	118	3909	1483	1075	0	6585
Potato	3430	29,367	19,509	742	63	53,110
Tomato	20,315	39,235	19,715	9667	892	89,825
Cowpea	0	0	0	0	516	516
Sugarcane	0	0	0	21,498	113,158	134,656
Fruits	7630	271,445	195,456	49,340	892	524,763
Others	130,215	170,803	158,604	109,379	113,329	682,329
Total	333,729	1,077,938	828,141	548,099	262,932	3,050,840
Grand total	712,005	2,159,665	1,783,669	1,120,787	520,818	6,296,944
CI (%)	188	200	187	196	198	194

this category. Table 9.1 indicated that the highest cropped area was found in the second agro-climatic zone, namely 2,159,665 ha, where the highest cropping index value existed. Furthermore, the lowest value of cropping index existed in the third agro-climatic zone. The average value of cropping index over Egypt was 194%.

Cropping Index for Cropping Pattern that Increases Food Security

The suggested cropping pattern that can increase food security included implementing intercropping systems and cultivation of three crops per year. We assessed the effect of intercropping wheat on winter tomato and on sugar beet. Relay intercropping cotton on wheat, as well as intercropping wheat with sugarcane and under fruit trees, was also assessed. Faba bean intercropped with sugar beet and winter tomato was also evaluated. Furthermore, intercropping faba bean with sugarcane and under fruit trees was also assessed. Maize intercropped with soybean, summer tomato, peanut, and sorghum was evaluated. Moreover, sunflower intercropped with soybean, summer tomato, sugarcane, and fruit trees was assessed. Finally, intercropping cowpea with maize and intercropping cowpea with sunflower was also evaluated.

Table 9.2 Total cultivated area and cropping index (CI, %) in the agro-climatic zones of Egypt under the increasing food security cropping pattern

	Zone 1	Zone 2	Zone 3	Zone 4	Zone 5	Total
Winter crops						
Wheat	176,915	505,182	462,478	366,127	114,581	1,625,284
Faba bean	27,067	93,337	63,038	34,846	26,437	244,724
Clover	58,505	240,800	201,591	110,888	12,956	624,740
Short season	100,000	100,000	100,000	100,000	100,000	500,000
Onion	18,225	57,015	39,075	31,711	2147	148,173
Tomato	2452	17,972	29,525	13,489	6736	70,173
Potato	5033	57,205	30,144	9674	230	102,285
Sugar beet	68,094	88,990	47,591	26,517	0	231,193
Others	87,344	198,881	213,357	35,285	145,325	680,192
Total	543,634	1,359,383	1,186,798	728,537	408,412	4,226,764
Summer crops						
Cotton	31,381	43,829	22,105	3033	0	100,349
Rice	103,853	304,770	97,627	0	0	506,249
Maize	45,715	233,440	388,459	431,470	43,596	1,142,680
Soybean	430	847	125	12,728	0	14,130
Sunflower	10,808	315,436	11,799	12,107	0	350,150
Early sunflower	8333	8333	8333	8333	8333	41,665
Potato	3430	29,367	19,509	742	63	53,110
Tomato	20,315	39,235	19,715	9667	892	89,825
Cowpea	25,533	152,350	220,500	239,198	24,375	661,955
Sugarcane	0	0	0	21,498	113,158	134,656
Fruits	7630	271,445	195,456	35,353	892	510,776
Others	130,215	170,803	158,604	68,562	113,158	641,341
Total	387,643	1,569,854	1,142,233	842,691	304,465	4,246,886
Grand total	931,277	2,929,237	2,329,031	1,571,228	712,877	8,473,649
CI (%)	246	271	244	274	271	261

Cultivation of three crops per year included cultivation of short-season clover as an early winter crop between winter and summer crop and cultivation of sunflower as an early summer crop between summer and winter crops.

The highest and the lowest cultivated area under this cropping pattern was found in the second agro-climatic zone and the fourth agro-climatic zone, with cropping index value equal to 274%. The lowest value of cropping index was found to be equal to 244%, whereas the average value of cropping index over Egypt was found to be 259% (Table 9.2). Table 9.2 also indicated that in the first agro-climatic zone, cropping index was 246%, whereas it was 188% under prevailing cropping pattern (Table 9.1), with 58% increase. This result can be attributed to the implementation of polycropping, taking into consideration that no extra irrigation water was needed to implement it because cultivation on raised beds resulted in 20% saving in the applied irrigation water. In the second agro-climatic zone, an increase of 71% was attained in the value of cropping index, in comparison with its counterpart value in the prevailing cropping pattern, whereas, in the third, fourth, and fifth agro-climatic zones, the percentage of increase in the value of cropping index was 57, 78, and 73%, respectively, compared to the prevailing cropping pattern. Thus, the highest percentage of increase in the value of cropping index was found in the fourth agro-climatic zone, as a result of increase in the cultivated area of wheat and maize, cowpea, and sunflower through intercropping, more than the other zones.

Cropping Index for Cropping Pattern that Faces Water Scarcity

To face the negative effect of water scarcity, we suggested to implement cultivation on raised beds to save 20% of the applied water to surface irrigation, which is the prevailing system in Egypt. Furthermore, the above-mentioned intercropping systems and cultivation of three crops per year will also be implemented. The second agro-climatic zone attained the highest cultivated area under this cropping pattern, where the highest value of cropping index was obtained. The fifth agro-climatic zone attained the lowest cultivated area under this cropping pattern. The lowest value of cropping index was found in the first agro-climatic zone (Table 9.3).

The percentage of increase in the value of cropping index in the first agro-climatic zone was 82%, compared to its counterpart value in the prevailing cropping pattern, whereas this value was 107% in the second agro-climatic zone and was 110% in the third agro-climatic zone. In the fourth and fifth agro-climatic zones, the percentage of increase in the value of cropping index was 98 and 94%, where the highest increase was in the fourth zone. This high percentage resulted in the increase in maize, soybean, and cowpea cultivated area more than the other zones (Table 9.3).

Table 9.3 Total cultivated area and cropping index (CI, %) in the agro-climatic zones of Egypt under the cropping pattern that faces water scarcity

	Zone 1	Zone 2	Zone 3	Zone 4	Zone 5	Total
Winter crops						
Wheat	203,952	594,137	560,177	446,379	126,498	1,931,144
Faba bean	28,681	95,603	64,134	27,310	26,500	242,228
Clover	71,475	294,906	249,825	137,067	15,262	768,535
Short season	100,000	100,000	100,000	100,000	100,000	500,000
Onion	22,080	67,251	45,615	36,578	2287	173,811
Tomato	2452	17,972	29,525	13,489	6736	70,173
Potato	6103	68,811	38,537	12,394	230	126,075
Sugar beet	68,094	88,990	47,591	26,517	0	231,193
Others	95,570	209,410	291,802	40,293	145,401	782,475
Total	598,406	1,537,079	1,427,205	840,027	422,916	4,825,633
Summer crops						
Cotton	31,381	43,829	22,105	3033	0	100,349
Rice	132,792	384,104	122,432	0	0	639,328
Maize	45,714	232,967	343,746	380,049	40,661	1,043,137
Soybean	430	847	125	12,728	0	14,130
Sunflower	10,808	314,758	11,799	12,107	18,571	368,043
Early sunflower	8333	8333	8333	8333	8333	41,665
Potato	3491	36,055	24,919	742	75	65,282
Tomato	20,315	39,235	19,715	9667	892	89,825
Cowpea	25,533	152,350	220,500	239,198	24,375	661,955
Sugarcane	0	0	0	21,498	113,158	134,656
Fruits	7630	271,445	195,456	49,340	25,003	548,874
Others	136,280	296,219	444,245	109,361	113,328	1,099,432
Total	422,708	1,780,142	1,413,376	846,057	344,394	4,806,676
Grand total	1,021,115	3,317,221	2,840,581	1,686,084	767,309	9,632,309
CI (%)	270	307	297	294	292	297

Cropping Index for Cropping Pattern that Faces Salinity Stress

In Egypt, salinity exists in 830,000 ha located in the first, second, and third agro-climatic zones. This salt-affected area is distributed by 36, 23, and 8% in these three agro-climatic zones, respectively. Thus, we suggested implementing one crop rotation for soils with high salinity level, one crop rotation for soils with medium salinity level and another crop rotation for soils with low salinity level. Cultivation

Table 9.4 Total cultivated area and cropping index (CI, %) in the agro-climatic zones of Egypt under the cropping pattern that faces salinity stress

	Zone 1	Zone 2	Zone 3	Zone 4	Zone 5	Total
Winter crops						
Wheat	177,699	549,989	546,494	446,379	126,498	1,847,060
Faba bean	21,090	77,909	59,412	27,310	26,500	212,221
Clover	66,806	282,461	245,966	137,067	15,262	747,562
Short season	64,000	77,000	92,000	100,000	100,000	433,000
Onion	20,692	64,897	45,092	36,578	2287	169,546
Tomato	2452	17,972	29,525	13,489	6736	70,173
Potato	5717	66,141	37,865	12,394	230	122,349
Sugar beet	68,094	88,990	47,591	26,517	0	231,193
Others	92,608	206,988	285,526	40,293	145,401	770,817
Total	519,159	1,432,348	1,389,471	840,027	422,916	4,603,921
Summer crops						
Cotton	31,381	43,829	22,105	3033	0	100,349
Rice	122,374	365,857	120,448	0	0	608,678
Maize	42,346	228,543	341,328	380,049	40,661	1,032,927
Soybean	430	847	125	12,728	0	14,130
Sunflower	6960	243,263	10,974	12,107	18,571	291,874
Early sunflower	5333	6416	7666	8333	8333	36,082
Potato	3469	34,517	24,486	742	75	63,289
Tomato	20,315	39,235	19,715	9667	892	89,825
Cowpea	16,341	117,309	202,860	239,198	24,375	600,083
Sugarcane	0	0	0	21,498	113,158	134,656
Fruits	7630	271,445	195,456	49,340	25,003	548,874
Others	134,096	290,373	429,393	109,361	113,328	1,076,552
Total	390,676	1,641,635	1,374,557	846,057	344,394	4,597,318
Grand total	909,835	3,073,983	2,764,028	1,686,084	767,309	9,201,238
CI (%)	241	284	289	294	292	283

will be done on raised beds to save the applied irrigation water and use the saved amount to satisfy leaching requirements.

In addition, we suggested implementing intercropping systems and cultivation of three crops per year (polycropping). Table 9.4 indicated that the total cultivated area was the highest in the second agro-climatic zone and was the lowest in the fifth agro-climatic zone. However, the highest cropping index value was found in the fourth agro-climatic zone and the lowest value was found in the first agro-climatic zone. The average value of cropping index over the five agro-climatic zones was 283%.

The percentage of increase in the value of cropping index in the first agro-climatic zone was 53%, compared to its counterpart value in the prevailing cropping pattern, whereas this value was 84% in the second agro-climatic zone and was 102% in the third agro-climatic zone. In the fourth and fifth agro-climatic zones, the percentage of increase in the value of cropping index was 98 and 94%, compared to its counterpart value in the prevailing cropping pattern. It worth noting that the lowest value of cropping index in the first agro-climatic zone was due to the high salinity level in this zone. The highest value of cropping index was found in the fourth agro-climatic zone.

Table 9.5 Total cultivated area and cropping index (CI, %) in the agro-climatic zones of Egypt under the cropping pattern that faces climate change stress

	Zone 1	Zone 2	Zone 3	Zone 4	Zone 5	Total
Winter crops						
Wheat	187,695	545,557	516,338	413,715	110,095	1,773,400
Faba bean	26,461	84,465	56,313	24,157	22,411	213,807
Clover	65,042	295,414	227,340	124,731	13,889	726,417
Short season	91,000	91,000	91,000	91,000	91,000	455,000
Onion	19,210	60,333	39,685	31,823	1990	153,041
Tomato	2256	3870	27,163	12,410	6197	51,895
Potato	5736	60,328	36,225	11,651	217	114,156
Sugar beet	62,647	65,568	43,784	24,395	0	196,394
Others	86,013	52,115	262,622	36,264	131,313	568,326
Total	546,059	1,258,650	1,300,469	770,145	377,112	4,252,435
Summer crops						
Cotton	27,302	41,664	19,232	2639	0	90,836
Rice	116,857	386,031	107,740	0	0	610,628
Maize	36,942	192,213	278,773	308,497	33,055	849,480
Soybean	357	481	104	10,564	0	11,505
Sunflower	8697	258,215	9527	9918	15,402	301,759
Early sunflower	6917	6917	6917	6917	6917	34,584
Potato	2933	31,066	20,932	623	53	55,606
Tomato	16,252	11,157	15,772	7734	713	51,628
Cowpea	20,683	123,458	178,626	193,766	19,325	661,439
Sugarcane	0	0	0	17,844	93,921	111,765
Fruits	6333	60,410	162,228	40,953	20,752	290,677
Others	114,475	225,928	371,319	91,864	32,032	898,780
Total	357,746	1,337,540	1,171,170	691,317	222,170	3,968,687
Grand total	903,805	2,596,190	2,471,639	1,461,463	599,282	8,220,671
CI (%)	239	240	259	255	252	253

Cropping Index for Cropping Pattern that Faces Climate Change Stress

Climate change is expected to increase evapotranspiration, kc values, and consequently water requirements for the cultivated crops in the five agro-climatic zones in Egypt. As a result, the cultivated area will be decreased.

Table 9.5 indicated that the total cultivated area was the highest in the second agro-climatic zone and was the lowest in the fifth agro-climatic zone. The lowest value of cropping index was found in the first agro-climatic zone and the highest value was found in the third agro-climatic zone, namely 259%.

The value of cropping index under this cropping pattern will be the lowest, compared to all studied cropping patterns as a result of abiotic stress that climate change will do. The percentage of increase in cropping index in the first and second agro-climatic zones was 51 and 40%, compared to its counterpart value in the prevailing cropping pattern. In the third, fourth, and fifth agro-climatic zones, the values were 72, 59, and 54%, respectively, where the fourth zone attained the highest value of cropping index.

Conclusion

In conclusion, the highest cultivated area was found in the second agro-climatic zone and the lowest cultivated area was found in the fifth agro-climatic zone. The highest value of cropping index was found in the second agro-climatic zone under prevailing cropping pattern, cropping pattern that increases food security, and cropping pattern that faces water scarcity. Furthermore, in the cropping pattern that faces salinity and cropping pattern that faces climate change, the highest value of cropping index was found in the third agro-climatic zone. The highest percentage of increase in the cropping index value was found in the fourth agro-climatic zone.

References

Madari DM, Shekadar SI (2015) Impact of irrigation on cropping pattern and production with special reference to Vijaour district. Golden Res Thoughts 4(8):320–325

Ouda S, Noreldin T (2017) Evapotranspiration data to determine agro-climatic zones in Egypt. J Water Land Dev 32(I–III):79–86

Shafshak SE, El-Debaby AA (1975) Crop rotation and agricultural intensification. College of Agriculture, Zagazig University Press